Alignment To Li

Alignment To Light
The Original Transcripts

Channelled by Julie Soskin

Ashgrove Press, Bath

Published in Great Britain by
ASHGROVE PRESS LIMITED
7 Locksbrook Road Estate
Bath BA1 3DZ

First Ashgrove Press edition 1994

ISBN 1-85398-074-9

Printed and bound in the United Kingdom by
The Cromwell Press, Broughton Gifford, Wilts

CONTENTS

PUBLISHER'S NOTE

This "Original Edition" is taken from the unedited transcripts of Julie Soskin's channelling, and is reproduced here verbatim in its entirety, save for corrections made by the author.

The first two books in Julie's trilogy, "The Wind of Change" and "The Cosmic Dance" (both available from Ashgrove Press), were edited in a style which rendered them easily accessible to a wide readership, including those who were previously unfamiliar with channelled material.

It was the Author's wish, however, that "Alignment to Light", should be published in its original form, as this would retain the exact meaning and essential poetry of the work.

ACKNOWLEDGEMENTS

My grateful thanks go to: Jane and David, Dee for her organisation, Jenny for her Light, Gwen who sat with me throughout and Rupert without whom I could not do half as much. But most of all to those that see the challenge and dare to leap.

I wish you all joy

Julie

1. CLEANSING

There is a stillness entering the hearts and souls of man. We welcome you and we give thanks for this occasion to speak with you.

Man has journeyed very far in his myriad experiences. He has grown, he has been tortured and he has suffered, but the torturing and the suffering was mans' own idea, it was never God's. It was, in your words, an experiment. An experiment in material manifestation for the growth and the wellbeing, not just of this Earth planet, but for all the cosmos, because the experiences of man have helped the energy forces of all.

This is the time of great change and great transition never known before in mans' history, and it is only possible that man can exist because of the progress that he has made. He has generally freed himself from the karmic influences, and as we speak to you in your time, your year of 1992, we can say to you that you are already in the changing period. It is no longer some nebulous idea for the future, it is here today - now. Sensitive ones have recognised this. You have worked on yourselves from the physical, psychological, mental and emotional levels to create a toughness within. This rock-like energy within allows you to shake off your cloak of karma, and many of you are experiencing what you would call a 'new age' and 'clean age' in your lives.

There is a process where you walk through a doorway. This has been referred to as 'initiation' in the past. As you walk through this doorway into Light, you will throw off the cloak of karma and you will then have a clean page on which to work. We have spoken of this before. We have also spoken of the necessity to clear the astral planes and the astral body of its negativity. As well as all the good thoughts, meditations, heightened consciousness work that is being done on the Earth, there is as much work being done on the astral planes. Many of you have access to a group energy on the higher level of the astral. This, some of you refer to as the master ray. This master ray, or the teachers, or professors, whatever you want to call them, are a group that can communicate easily to you, and are here for your benefit. They have a deva counterpart. Their deva counterpart is what you can refer to as the angelic beings of Light. These angelic beings help with the energy, while the professors, the masters, help with the intuition, the wisdom and the knowledge.

One or two of you here have recognised there is a point in man's growth where he ceases to want to know, because he is so complete with the whole that specific information is no longer necessary for his intellect. He just becomes. He just is. This is the stillness to which we refer - the stillness of knowing that all is well no matter what is going on outside yourselves. All is well, dear friends of Light. We say to you very clearly today, all is very well!

It is not strictly true to speak of this planet as an experiment. That makes it sound harsh, and it is not so. But if we were, for a moment, to speak of it as an experiment, we could truly say to you that the experiment has worked and has excelled any expectations. This may perplex you when you still see trouble in your world, when you are aware of violence and suffering. This violence and suffering is coming up to the surface to be dispelled. Only by its revelation can it be dissipated. You will see, unfortunately, some pockets of very violent situations this year, but they will only be pockets and they will not even affect the near vicinity. They are being, as it were, collected in a pile, just as you collect your dust into a little pile before you clear it. This particular kind of negativity is the kind that has been affected over the years by the astral negativity, and only by it coming into pockets can it be cleared, up to and through the astral. The astral plane is not a separate dimension, but is a dimension that has always overlapped your own physical being. So when you speak of the spirit being above you, you are actually incorrect, because the spirit is all around you, is with you. As man has grown he has been able to see and hear and perceive the astral forces. It is within those astral forces that you have, what some of you still refer to as, 'guides'. There has never been a time where man has been alone, never, nor will there ever be.

Many of you want to know the future. Some of you still have some reticence about the information and about the feelings that you have of the violence and the trouble in store. It would be untrue of us to say that there would be no trouble, because there will. We ask you again, when it occurs, when you see it in the media, around you, do not dwell upon it. Put it in Light. Circle it in your mind in Light. All is being worked through. Of every description there has been of this period, the most accurate is the one that says it is the clearing time. The clearing time is here, and when it is finished, which in your terms is a very short period of time, man will be so different in consciousness he will not even be asking the same questions you ask today. That is not so easy for you to imagine, but as perception and consciousness alters, man will have far less need to want to know.

Again we speak of a state of being. This means that your attitude to life, the things that you emanate, are very different to the ones that you emanate today. We ask you now to open wide your mind and imagine, just for one moment, yourselves without fear in a state of perfected peace. With that state of perfected peace it does not matter what is going to become of you, where you are going to be, what job you have, what relationship you are in, because you know that the peace, the stillness and complete happiness is permanent. Some of you ask "Well this is wonderful, but what does it mean in aspect to striving? Man surely need trouble? He needs roughness to climb and pull forward?". Yes, indeed that has been the truth. But your whole planet is transmuting to be a planet of Light. So it will cease to be a karmic planet, and the sort of experiences that need that roughness, that little bit of suffering and anxiety to allow that soul to climb, will go on in other areas of the cosmos. So if a soul needs to reincarnate in a realm where karma exists, it will do so. This does not mean that the progression of the Earth will stagnate, far from it. Your Earth will be a permeation of Light force energy. It will be a leading and a healing energy. It is very easy to touch your stillness within. When you get to the point where you want to, when you have dissolved your own negativities within, it is a tiny, tiny step through the doorway.

Many of you are fascinated with what the changes means in physical terms. Evolutionary process is in operation, and because the Earth is not only transmuting spiritually, but is going through a physical change. That physical change will, in itself, alter man. The organs of man will be less heavy, less dense. One or two of your organs will disappear. Your mouth, your jaw will alter, and your blood will be slightly different. But all this is happening over a period of years. Some of you have experienced the cells in your body 'bubbling up', much like the molecules in a kettle when it boils. Your cells are truly being stirred by the energy that is coming into the planet. August 1991 was the first time that, because of a clearing already achieved and the Earth's position to the energy of Light, the energy permeated and was felt in several areas of your globe. The area closest to where you are now was in Ireland, South Ireland. That ray was able to penetrate because of the work done to clear the astral planes, which is the emotional body that is gradually disappearing from this planet. This ray is permeating and is still spreading over your continent. As it permeates through areas that are not so mature as this one, you will be seeing difficulties and, indeed, you will be seeing violence. Again, we urge you to remember this is to do with cleansing. The energy is not creating violence. The energy is pulling it up to deal with it, to dispel it. This energy, in your terms, could be called a blue ray, and some has been perceived as such. This blue ray will be in operation now for some years to come, until your planet finds its own physical place in your solar system.

We wish to take your minds above the Earth. For a moment imagine yourself out of the Earth's orbit, above the solar system, looking out around you. If you do so, you realise the constant, permanent movement of forces in the galaxy and the cosmos. New suns are being born frequently. Old suns are dying. There is a constant flow of change. Your planet is going through a cosmic change, and the positioning of it in its solar system and in relation to the cosmos is altering. It is momentary going back to where it was thousands of years ago, and then it will move on again. This is why so many of you here at this time are old souls, souls that remember or are beginning to remember your early times on this planet. There is a connection with the stars. A star that is not your own sun has played a major role in you development and growth. But there has been more than one star connected to you throughout history. Now, as the movement carries on, your Earth will literally be in the same position compared to the star that it was thousands and thousands of years ago. This will again align the energy force, but it won't be for very long because it has a new movement to take.

We ask you again to take your minds above your planet. You must think in cosmic terms to begin to understand what we are referring. The next time there is a clear sky at night spend a few moments looking up to the stars. Realise that those stars are not even a tiny fraction of the whole. When you think on that you will truly be awe inspired.

Your planet will survive. It is not being destroyed. It is a joyous transmutation of Light force. And to those people who greet this information with fear we say, dear friends of Light, without change you would not be here today.

Man, alive at this time, is truly fortunate. He is going to experience the transcendental movement of time, and there will be a period, approximately, in your terms seven years, where time will be a difficult dimension for you. Some of you will feel lost in time, dissociated with the Earth plane, and we have already spoken about some of the reasons for this disassociation, but it is also to do with the movement of your planet around the sun. Understanding that you measure your linear time by the sun, you can then well understand that if the movement of the sun and the planets alter, your feeling of time will alter too. This is really an opportunity for some of you to time travel! This may sound, to some of you, a little far fetched. You have in your minds thought of time travel, perhaps, in a machine that would be invented one day. But time travel does not need a machine. It is, in your terms, another dimension. As the dimensions are loosening at this time (which is why you can associate yourself with the astral plane, which is in itself a dimension) you will also be able to acknowledge the swift change and the diving into other times. You ask "How can we possibly keep a bearing on our own lives, our families and our friends?". We give you the advice to live in the moment, dear children of Light. Each moment perfected will create a fluidity, a fluidity that you will easily be able to walk through.

For some time you will have to still consider the material aspects, your work, your homes. You will see in the next two or three years the material aspects of the world breaking up. Again, this is a transmutation. But those of you who live without fear will not suffer through lack of material goods. What you need to do for your own survival, for your own comfort, you will do. If you need frivolity, dear children, enjoy yourselves. When you are attuned to Light everything you want is right, because you will not want anything that is wrong for you. Indeed, the spiritual forces have been close to you, to help you immerse in enjoyment and happiness. Happiness, of course, is, as you know, a by-product of correct thought and right living. This is not a pompous statement. It is not for you to feel you have to be the hermit. Embrace life in all its myriad experiences. There is no doubt, no fear, when you are aligned with Light. Most of all we have tried, throughout these communications with this particular channeller and others, to impress upon your world the deep necessity for losing your fear. Although a considerable amount of work had been done, there is still some loose ends on this matter of which we will now speak.

For many now the psychological approach is no longer necessary or applicable. Indeed, for some people it will create more of a negative flow. It has been also an excuse, an excuse not to go back to your own individual Light and your own knowledge - remembering, dear children, that you all have knowledge within. You do not need others to tell you what you know within your hearts and souls. This marvellous and wonderfully spiritually creative time is about man standing as one with God, and for that each individual must open up and accept his or her own power. Of course there are people who have had troubled childhoods, or who have had trouble in past lives - indeed all of you have had trouble in past lives. We ask you not to concentrate your efforts on delving into the past.

The energies coming now, of which the blue ray is one, will help you. You can and you will be complete within yourselves, by knowing and the touching of your Light force within. Of course, we are not saying to you that you do not need friends or advisers. Listen to your advisers, listen to your friends, but make up your own mind. It does not matter how much you trust one other person. We ask you even with the information coming through this channeller, trust your own heart, your own feelings, because it is through that, that man will truly transcend into the being of Light.

This means that some of you, who have dedicated your life to some form of therapy, will have to change or bring in your own new feelings to that particular therapy. You will do this instinctively and intuitively. We do not say that no therapy is right. For some there still will be a need for those psychological type of therapies. But the vast majority of people are beyond that now. Indeed, we are very glad to report that in a few short years the need for therapies, as you have known them, will diminish entirely. Think on this and you will begin to see and know the possibility of which we speak.

Every single man has an opportunity to be at one with God. Every single man. Think on that too, dear children - none of you are exempt. As each individual races to the level of alignment with Light, it adds to the whole, it strengthens the healing force. That healing force permeates from you and is creating a ray of Light high in your Earth's atmosphere. If you could perceive it in your terms it would be rather like a bubble that is getting bigger and bigger. Within that bubble is the clarity beyond which you can comprehend. This clarity will not allow negative forces. It will not be possible for evil to permeate through that bubble of Light, which you have created by your experience and by your work on yourselves, not just in this lifetime, but in all those other ones. You have created the stillness. Some of you who have thought about your past have assumed that you got it wrong somewhere. Some of you have assumed that you must have misused this power. Most of you did not misuse it. The majority of you elected to go through the suffering, abandoning power. This was a kind of sacrifice, but a very willing and knowing one, because it was known, over many thousands of years ago, that man would reach this point now, or at least he had the potential to do so. He has reached this potential now.

We speak of the physical body and how it has shaken. Indeed, many of you have experienced, recently, rather disturbing feelings in your body. This you have assumed automatically to be the normal coughs, colds and viruses that you all know about. The doctors speak of a new strain of virus, and in some ways they are accurate. But what has occurred is that the purity of the rays coming through are working down to the many bodies of man, from the highest, through to the mental, the emotional, and lastly, as always, into the physical. These physical manifestations are not because you are doing something wrong, they are because you have accommodated the new energy flow. So some of you have mistakenly thought "I am ill and it must be because I am out of balance. It must be because I have a disturbing thought, or I haven't cleared this or that". It is not so. Most of these physical disturbances are the lower manifestations of the new forces. As the body accommodates them it will adjust, some of you very quickly, some of you a little slower. It is not in any way life threatening, but you will feel the sensation of this disorientation, as a physical heaviness of the body accommodates the force of Light. Again, we urge you to realise this is not anything to fear, but is everything to be joyous of. We spoke before of birth pains, and this is what you can define these physical manifestations - as birth pains and the joyousness of what is connected to the birth of Light in man.

There is much more that we are going to say, but in some ways there is nothing left to say, but again, we repeat, those of you who have walked through the doorway of Light understand. What is physical death if not a manifestation, a birth into spirit. Take away your fear of death and you will see the perspective of what is happening.

As some of you are teetering on the brink of this awakening, and particularly in your cultural energy of the British Isles, you have what you describe as an idiosyncrasy where you feel almost justified in appearing servile, not being forthright and assertive. This means that you do not very readily like the idea of power, and those people that have been bold enough to state their power, you have laughed at. You have said in you minds "Who are they to say that!". But we say to you now, each and every one of you has the opportunity of taking hold of the rod of power. You would not be at that state if it were not right. And it is right, because you are coming into your acknowledgement of all and everything. Some of you like to call it higher consciousness, higher self, spirit, God, Atman. Call it what you will, but it is a part of you and it is a part of everyone.

There are always difficulties at times of change. And with a change so radical as this you would not expect there not to be some disturbances. Keep your own stillness. Keep calm and still within, even at the most testing of times. There will be a situation coming that you will not intellectually comprehend. So be it. Your connection with the force of Light is what is going to allow you to move forward and transcend. We urge you, and we give you a surge of energy to help in your enlightenment. The guardians of your planet who have worked so strenuously are here again. They were here at the very beginning of your Earth's manifestation. They periodically come close and they are close now. So you see, dear children of Light, none of you are alone. But it is within your own stillness, your own quietness, your own connection to Light that will allow you to transcend.

We ask you for a few moments to close your eyes and absorb the ray of Light. These little surges of energy are a little start, a little spring-board, because once you feel and acknowledge without fear these forces, you can never go backwards. You cannot, after all, unlearn knowledge, and these energy forces are like knowledge to the body. You can never go backwards. There is only one direction and that is forward, into Light.

Some of you are wondering how this process of channelling works. In the most simplest of ways we speak of an alignment of Light forces. It is possible for every single one of you to accommodate channelling, but it is easier for you, perhaps, to contact the group energy of the master ray at this time. Because linking and touching the group energy, the master ray, still has a very human faculty, and so it relates very well to your thought processes. The channelling today is one that can only speak in terms of the higher Light, the higher information. It does not work so well for the personal consideration although, dear friends of Light, within your own soul and your own connection with the Light forces, you have all the answers.

We will end today in a simple way because very often the simple words are best. We could have chosen to talk in very complicated manners, and for some channelling and channellers that is occurring. Simple resonance of truth is the easiest to understand and the easiest to acknowledge. Each little cell, each little bit of skin on your body vibrates exactly the same way as the whole cosmos. You could almost say that within your body are many worlds, thousands and millions of worlds, wheels within wheels. We can not tell you to trust, but trust is a necessity. We could give you exercises, meditations. Some of these will help - do help. But the clearest simplest advice that we can say to you is that you are free, you are at one. Absorb that, know that, and within that knowledge will come trust and strength. We give you all the blessings. We give you all the elements of joy.

II. THE HEALING PLANET

Humanity is half way up the mountain now, and although some of the journey is still perilous, mankind is used to the terrain. The sands of time are shifting very quickly, and it is as if your time has accelerated. So man is experiencing his problems, his suffering, his pain in a very intense way at this point. It is no longer the point of departure, because mankind has entered this new space and realm. It truly is a transitionary period, unlike anything ever encountered.

Some of you draw parallels of past times, but it is not even as it was then. The dimensions are melting into one. Time and space are also melting into one. It is as if outer time and inner time are not in unison. In inner time you are, as it were, in slow motion, and outer time is speeding up. This is the best way to describe the difference of the turning movement of the planets in your solar system and others. The outer cosmic wheel is shifting, is turning faster. In relation to the other planets and stars you still feel you are in the same space. But there has been much movement not yet discovered. You have always been, and always will be, affected by the outer energies on your planet. Now some of these outer energies are closer. The atmosphere around your globe has thinned, and as it has thinned, so the Light and the emanation of the energy from other stars permeates through and is affecting your globe very strongly. This occurred for you last year, it is still in operation and has spread very quickly. With this new energy force that comes from the stars, the dimensions are melting where it is physically touching your globe. It is now going to be quite difficult for those men of logic to keep hold of their structures and formats. They will have to move with the times. They will have to 'spread out' their brains and their minds to accommodate the speeded up forces. This all seems to some of you quite untouched by reality, and many of you are trying to hold onto the past realities. We say to you this; When you are on dry land you can walk freely. But when you are in the water you cannot walk, you must swim. And so it is now, being in a different space, in a different element almost, that you have to do things differently. You have to swim with the current, the energy in which you find yourself. Man is still frantically looking for influences, for something to 'hang his hat upon', a force, a power, proof, some kind of conscious realisation on a physical level. It will not be long before man sees enormous changes physically.

The mind of man has always been devious. It always sees what it wants. And all the while things are moving and changing around. Some people are putting their own understandable connotations on what they see. It is still unfortunately only a very few that are seeing even some of the picture. However, the favourable, valuable thing that is occurring is that more people are coming into their own conscious awareness of the further dimensions. It is from that they will understand and know what needs to be done. We have spoken a lot about the movement of energy, the ebbing and flowing, the 'breathing out' and the 'breathing in', and we hope by now that this is understood on whatever level it can be understood.

Your globe is shrinking, is breathing in, so to speak. The magnetic force within its orb is pulling. Its shape is subtly moving. This, as yet, has not had too many dynamic occurrences. But as it shifts and moves it is obviously going to shift and move the outer crust, meaning that there will be some movement of land mass. This is already happening to some degree, but will happen more as we move on in time. It is in a sense 'wobbling' on its axis. As science has known for many a long year, your planet does wobble, and every so often that wobble creates a reverberation that alters the trajectory of the planet. This, now, is more than an automatic wobble. It is from the magnetic core that it is shifting and moving. With the wobble that is constantly there, it will create a dynamic force. There will eventually be a complete change of axis pole. This is why you, in human terms, are having so many experiences of the male and female energies. All this is coming to the surface because the polarity of your own planet is moving, is shifting, so that the male and female counterpart in the human body, the psyche, is coming into consciousness. You are asking yourselves about male and female energies. You are working with those forces. Everything has its equal in the Universe. Everything is affected by every other thing. As the poles in your planet move, so, in your own bodily functions, do your poles - your masculine, your feminine, your positive, your negative, move, change and come into consciousness.

What is very strongly coming into consciousness now is mans' power. When we speak of man, we mean, of course, man and woman. Mankind's power is coming into consciousness in the same way that the Earth's power and energy is coming into its consciousness. Because, unbelievable though it may seem to some of you, your planet has a consciousness of its own. You have for too long considered the physical planet something like dead wood. The physical planet is not dead wood, it is a moveable conscious force, and each planet, each orb has its consciousness, its note, its vibration, a vibrational quality, just as man himself has a note and a vibrational frequency that resonates out into the Universe. On one level we could speak to you in terms of sound - it is because of the frequency, the vibration of the note of each orb, keeping in tune with the rest of its system, and therefore with the whole cosmos. It is not sound as you understand sound, but it is a vibrational frequency that, if you had ears to hear, could hear a note, and the so-called music of the spheres is what keeps every planet, every orb in its position. It is connected with the gravitational pull of the planets. The note is the gravitational pull. Sometime in your future your planet and the conscious energy within your planet will begin to understand this and how it works.

There is going to be a mass exodus from your planet. We have spoken of this briefly before. This mass exodus is for many reasons, but the predominant one being that almost as much as three quarters of the souls present now on your planet will move on to other realms of experience. Some of these incarnating in different worlds, some karmic worlds and some beyond the karmic forces. This, as you know, has happened before. There was, from another planet, a mass exodus of souls. That mass exodus came to your planet Earth, and created an enormous push forward of evolution for Earth at that stage. It is not uncommon for this to occur when these shifts of the gravitational pull and the notes and the harmonics alter. It is almost obvious that many souls will part at this time. They will then re-manifest in many different ways, in many different guises. There is a younger planet than your own that is in great need of the mass exodus of energy. It is a younger and vibrant planet, not unlike your own a few million years ago, and it needs the extra injection of forces. There are those that can accommodate that in their next evolutionary process.

There is a never ending flow of energy. For many, many years man has thought only in terms of his own small sphere. He was only aware of his close vicinity. There were many years when man did not even know of the existence of other men in other places on his own globe. Think of the exodus from your country to other countries, the exploration, the adventure, as man went across the Atlantic and the Pacific, found other cultures and moved into new energy forces. This is exactly what is occurring now. There is a time when a culmination point, an outgoing force, seeps into consciousness, whereby whatever energy is in force, it moves out like iron filings from a magnet. The energy is like a star burst in the sky, and when it occurs it is joyous. It is like watching an enormous firework display, as the Lights from different areas fly out, accommodating and Lighting other globes and other places. It is in many ways an exciting time. Of course all these words of description that we give are objective words we merely use to paint you an image, a picture.

There will come a time when the energies, the souls that move on to newer, younger planets, will again have worked through and balanced and helped and grown, and they then in turn will move on again and again and again.

No one, no consciousness can say with absolute certainty what transmutations will take place within this 'jumping off' period. It is not just to your planet that this is happening. It is an alignment of forces. If you want, it is a conscious acknowledgement between different force fields that this kind of accommodation is needed and will be attracted. We speak now in what you define as metaphysical terms. But all is one within a flow of forces. You will see, within the next few months and years, others experiencing what you have experienced. You will see others come into their consciousness, their awareness, their realisation of where they have come from, and for some, where they are going. It is rather like the early pilgrims setting off on a ship across the Atlantic - anticipation, excitement, a little bit of apprehension, but willingness to start again in a cleaner fresher field. The cosmos is grateful for these souls of adventure, for these ones that make this leap without knowing where they will end. Like attracts like. Those souls that can stay will stay, accommodating the new forces on your planet that will, in a very short period of time, be very different.

We speak again of your planet being a fresh green land. It will be a regeneratory force for the cosmos. You know yourselves, those of you that are in touch with your spirituality, that you have what you define as a spiritual home, some place that you go, you touch, you feel attuned, you feel a part of the Universe, you feel healed, you feel refreshed. For some it is near water, on water. For others it is in forests, or trees. We say to you that your planet will be such a place for souls to come, to replenish their spirituality, their grace. Some of these souls will come unconsciously in a kind of sleep state, or at least that is the nearest words that you can understand. Some will come consciously, will live and exist for a while. You will find it difficult to imagine that your planet, that has been for so long a planet of strife and struggle, is becoming a healing place to accommodate those souls, who will need energy replenishment and growth. For those men and women who set sail across the Atlantic, they did not know for certain what was there. They knew that some would not arrive. But they knew they had to go. They needed to take their energy to a new place. So it is on a larger scale now. Your energy is needed in a different place, and so the travelling begins.

The mass exodus is starting now and will accelerate, particularly over the next few years. That is as far as we can project in your linear time. We could speak of the whole process taking one of your centuries, because a century from now it is most likely that your planet is settled into its healing energy force. But, as you well know, timescales are very difficult, because who can say when the wobble will set about its reverberation? It is not likely tomorrow, but it could be in a year or two's time, and that will create another force that accelerates this exodus. So you see, it is not so easy to give a time. This reverberation may not happen for ten or twenty years, although that is unlikely, because it is setting about a very strong vibration already.

Much of your healing work, consciously and unconsciously, has been done through your Earth's energy force. Even those that would abandon that idea automatically have worked with the Earth's energy. As the Earth reverberates it is not so easy, in some cases impossible, to work in the same way. This has left many healers feeling almost a failure in their work. But, of course, failure does not exist. Man must and will heal himself, so healers can only help the opening of the door. This will happen through the energy field of the healer being in a positive mode, and also through talking and emanating the energy of communication, which is still important for the patient. But it is largely through the energy field and the touch that will help, not from the Earth's energy; it can only help from the connection of the healer with his or her own force within, connected to the various cosmic rays that work through and align the patient. That aligning must in the last case be done by the individual. So it is only a reverberation of the energy field that the patient feels, holds onto, and uses sometimes, as a vast stick to jump over into their own consciousness. This is a little bit clumsy, the way that we are putting it to you. The healers are not able to contact the energy that they have in the past. It is only those healers that are absolutely strong within, that will truly be able to heal. This you are already seeing.

In many cases the healers have had to work furiously on themselves again, not in a psychological way, but with their energy and the maintaining of that energy within. This energy, of course, is still moving, so it is not even now fixed. How you go about your healing today will be different from next year. All we can say for your advice is that you must use your intuition and your own link with your inner core Light, your inner core knowledge and wisdom. When you do this you will not get it wrong. If any person tries to heal in a way that is no longer feasible, they will see it soon enough. But we emphasise, there is no such thing as failure.

There is no formula for Light workers. They made a conscious decision, in some cases many millions of years ago, to be here today. The way forward is assured. There is no possibility of stagnation. It is for Light workers as if they have pushed out into the cosmic realms, millions and millions of years ago - we speak, of course, in your terms. They have set about the motion of growth, of experience and of knowledge, and this has no end. There are very few healers that are not in this particular mould or space. Those that are not in this space are in a way charlatans to themselves and others. All those that can truly say that their desire has always been to grow are those that have already set about the motion of continuous movement forwards.

We know that from time to time some of you feel that you are on a plateau, you have stagnated, you are not perhaps doing enough. We say to you, categorically, there is never a time when you stop moving forward, never. These are times of great acceleration. There will not be many times when it feels slow, because largely you will still be aware of tremendous movement. For some of you, you have had an interim space over the past six to eight months where things have, in your terms, wobbled, and where you have fractured from your connection. This feeling of being parted or fractured from it was, as you well know, an illusion, and it was actually a part of the process of forward motion. This will take a little bit of time to acknowledge. It fits into a larger framework which connects to you being multi-dimensional beings. Future plus past memories will open to you in this process. This does not mean that you will even, in the short period of time of which we speak, have all knowledge at all times. You will be slipping in and out of your time. It is not possible for any living soul, particularly in a body, to acknowledge the whole aspect of the cosmos and the energy force. Quite literally, your mind can not take it. Your mind is opening to greater and greater thought patterns of frequencies, and that of course means that you will be aware of many, many things that you could not envisage even now.

From your perspective, here today, there is much ahead of you that you cannot even imagine. You can only see within your own framework. Your sound has its limitation, your sight its frequency limitations. There are things beyond your sight and that you know exist and you know within your soul that it exists, but you cannot physically see it. Therefore you can not make them logical and your mind cannot accept it. But as the frequency excels each way, above and below, in your terms - past and future, you will see more of the whole. But do not expect to see all.

The seven planes are melting into one. This will mean enormous revelations, awakening of consciousness of which you are already partaking. It will be like a flash of lightning consciousness, and in some cases for a split second you will sense and know all. We speak in terms of an analogy when we speak of a split second. Then you, whatever you are, whatever your frequency, will calm down to your own natural level, which undoubtedly will be beyond what was there before. It is very hard for us to put these things into words. The seven rays are connected with your planet frequencies. There are some beyond that. The seven rays have corresponded to man and to man's planet in the material way that they can acknowledge. But there are forces beyond the matter, the material. There are rays beyond the seven that you now understand. These are not necessarily, in your terms, above and below. They are different. You will also catch sight of the linear future as the energy force turns.

There is so much information that could be imparted at this time. We are having to make this over-simplistic because we very much need, at this time, to give this energy to man en masse. The rays of man are absorbing into his own sun. He is absorbing all those energy forces, creating a oneness never known to man before. We cannot expect you to fully comprehend. This is not a demeaning statement. When the brain is given too much information it implodes on itself; it is best to keep it simple.

We see in your hearts, goodness. We know your strength of purpose. There is no need to struggle with any of this information. We see the hearts of all men, and we can say to you, to bring comfort and warmth, in truth and reality, man's heart is very strong. The heart energy force is becoming large and vibrant. This is the force that will eventually lead to your planet being the planet of the heart, the planet of healing, the planet of strength.

III. PART OF GOD

The consciousness of man being raised means many things and has different effects on different living things, because the consciousness raised is not just in man, it is in every living thing. This of course includes the animals, the trees and plants. Consciousness is being aware of being. This has many levels; you can be aware of your physical body and there are living things that are not aware of their physical body. You are aware of pain, distress; to know what to do when you cut yourself. It is instinctual survival that is an awareness. It is a consciousness of a kind. The next step forward is being aware of your fellow beings. You can say "I am aware of myself", then you become aware of others. From that point you become aware of the energy that your fellow man emits whether it is on a physical level or other levels, and of course we come to the consciousness of the emotions - the emotional body. When man became conscious of his emotions this was a very big leap and indeed we have spoken about this in a previous book. The big leap that man made from the consciousness of the body to consciousness of emotions was a highly important leap for man, because in that evolution he then created an energy field, by his consciousness. He created an emotional body around the atmosphere of himself and your world. This atmosphere is the basis for your astral plane.

Then man has gradually become conscious of the mind and the psychology of the mind. This has created a mind energy, an intelligent force. Each time man has risen in the various consciousnesses by his ripple-like vibration emanating out from him, he was able to contact like energies. As there are not many globes like your globe, it was not possible to contact others of emotional or physical consciousness, but as you have come to the mind there is a universal, cosmic mind consciousness that can be tapped-into, and indeed is being tapped-into now.

We have missed out on a very important factor of evolution and that is the heart consciousness. The higher love vibration that is now becoming the base, the foundation for the human evolution to come. So it will no longer be the physical, the lowest vibration, it will be love as the lowest. Some of you will find this extraordinary because you have always anticipated that love was the highest frequency, and for many years that has been so. But we tell you now, it is not the highest frequency in the cosmos. It will now be the base for what is to come. This is the mind, the will, the being mass of completion. You are becoming increasingly aware of a consciousness beyond your own.

Be still, dear children of Light, be still. You can be aware of the gentle touch of a daisy in your hand and you can be aware of the mind of God, and even in that there is no end, there is no beginning. We are helping to send other frequencies of Light to seep into the consciousness of varying types of humans. It has been relatively easy to touch the sensitive, even though it has been hard for some of these sensitive intuitive people, because they have for many years been alone in their knowledge, they have been ridiculed, despised and denied. Now inspiration is coming to the mundane mind, to the mind of the intellectual, and to the mind of science. To begin with, the enlightenment could only seep through the most sensitive of the scientists, but in time it will seep through to all. It was important for man to climb up this way on his journey, in many cases blind to what was above, below and around him. It is only because of the work done that all men will have available at least some of the knowledge we impart.

Science is very close to the discovery of the fourth dimension, and although, in human terms, the fourth dimension is nearer, closer to you than the fifth, man is even closer to discovering this, the fifth dimension of Light and all that means. There are bodies of beings in communication with you now because the atmosphere has been opened. Some of you got a whisper of the information over the last three to five years particularly. A very few of you had a whisper of it prior to that. But now the energy is here, and you can use this energy, this Light force. It is like the sun permeating through the space in the clouds. It will come and go to start with, then the clouds will completely clear and you will be immersed in this energy of Light. It is a new atmosphere and it is affecting the matter, the spiritual, mental, the mind, the love, the heart vibrations, emotion and last, the physical. You will have to work on resolving your attachment to the physical and the emotional, and you will have to consciously detach from those lower frequencies to accommodate the Light vibration.

The heart energy, the love energy which is the physical manifestation of Light, is your base for the new generation of consciousness. The freeing of the fear from the lower centres creates possibilities beyond your imaginings. Your lives have been dictated by fear. They will no longer be, they cannot be. The sympathy and the compassion that you will feel for your fellow man is through the knowing of the pain and the frequency connection of the pain that they may feel. But as there will be vastly less pain and suffering, there will be, of course, less need for the compassion. You will work together as an harmonious force of Light. The Light beings that come are your future and they are your past. Like you, they have forgotten their history, and in coming close to you they remember. They dance on your laughter. They are immensely curious about how you have existed. They are beginning to recognise you as their children. We speak of "they" and we speak of "them" as though it is a body of people. It is not a body of people, it is a mass consciousness from other places in the cosmos. They come by your emittance of the higher frequencies. They come in answer to your call. They are Light bodies and they are a few steps above you in the evolutionary chain. You have been so bogged down by your globe and your physical sense of being that your globe has been isolated in the cosmos, isolated by your atmosphere - the atmosphere of emotions which encompass darkness.

Although it is factual to say that there is an equal and opposite of all things, it is not opposite as in the way you think of positive and negative. It is better to describe it as two sides to a coin and one side of your coin does not decrease in the value of the whole. We hope you understand this because some of you are looking for darkness in the cosmos. We have stated before and we will state again, there is virtually no darkness as you understand darkness in the cosmos. The majority of living vibrant planets are planets of Light. They have their polarity, but it is not polarity of darkness, and even when you speak of negative, you immediately think of something less important. You use the word negative as a subservient energy: "That person is negative, that force is negative". That thought and that feeling helped create a darkness around your globe. Think of two sides to a coin, equal in worth. That is polarity as the cosmos understands it, and when we use the word negative energy in the cosmos, we do not mean bad, black, evil, opposition, we simply mean different. There is matter and there is anti-matter. Anti-matter means invisible, means different from the physical. It is not evil force. You know very well that to create electricity you have your battery's positive and negative charge - one without the other does not create the energy. As so it is in the cosmos. You must put away finally and irrevocably the idea of dark atmospheres in the cosmos. Your thoughts, as they permeate out of your globe, are creating energy, and those thoughts must be clear. The cosmos will not allow destructive forces. You have only been allowed to work your destruction out on your planet, you will not be allowed to work it out in any other place. There will not be star wars, because the cosmos will repel those forces if they should enter above your atmosphere.

It is truly out of your frequency of thought to imagine and absorb the concept of being in touch with the whole consciousness of the cosmos. Even to the original thought processes, it is still vast in its being. The mind and the consciousness of God has many arms that draw in many forces, many beings, and each being that is drawn in is like a child being collected, nurtured and gathered in Light. You are our children because you are listening and aware of the mind of God.

We speak of strength, and just as we have described the many layers of consciousness, so it is with strength. You speak of physical strength muscles, you speak of emotional strength, the strength of love, then the strength of the intellect, the mind; we want you to realise now the strength of the whole of the source of God is at your disposal as you are absorbed into it. The intellect, of course, is the techniques of the mind. The mind is the vibration and the reflection of the mind energy of the whole of the source. "As above so below", you say; in reality there is not an above and below. All things are joined in unison, but let us say, as with God, so with man. Man is working out God's thought by his being in the physical body. Man is working out God's will. You speak of free will, and there is free will within the will of God. You are not controlled, but you cannot help being affected by the will of God, because your will is the physical clone of God's will. You have spoken much nonsense, dear children, when you speak of evil. Evil has been the working out of a destructive force, and it has created a strength unknown to many others in the Universe. You cannot possibly accommodate the reality of the knowledge of the whole, but you do from time to time see a reflection of it, and even from time to time you feel a connection to it.

We speak in your terms when we speak of time. Time is tied up with the loosening ties between the dimensions. Time is what holds the fragile cords together. The assumption of time is essential to the human condition. But as the physical and emotional conditions are altering therefore your relationship to time is also changing. Some of you that read these words are perplexed and do not understand and twist the meaning. We are aware with our last statement that some of you may consider this to mean that you will not have a physical body. As already spoken, you will indeed continue with a physical body but that, of course, is mutating to accommodate the frequency of Light, and the evolutionary process has a long way to go before you lose your body. The body will become for you a vessel of very much less importance. It will still have to be maintained, nurtured, fed, but the way that is done will be very different from the way that it has been done in the past. Remain still in your core and heart and soul. That stillness is imperative now. Without the still beingness you cannot evolve into what you must become, and that is a body of Light frequencies.

Do not be afraid, we are bringing you joyous information. Do not be afraid of your power. Do not be afraid of the inner strength which comes from the union of consciousness. Do not be afraid of the Light bodies from others worlds close to you. Do not be afraid because your Father is still with you. Rejoice in the coming of the new wave. Rejoice! The information that we bring is not about calamity, it is not about saying that in this time or that time this thing or that thing will happen. This information is to project your conscious mind into the very real possibilities, and we hope probabilities, of your growth. You have free will. Those that deny this information will have to wait for the next wheel to turn for their movement into Light, and the next wheel turning may take many, many Light years. It does not matter in terms of development, in terms of timelessness. But, Oh dear children, again we urge you to take this opportunity. There is no limit to souls that we can accommodate in the new Light. It is only your lack of trust that puts divisions on that. Even the highest consciousness on your planet, those of you that know much, never imagined you knew everything, because at best one individual only sees part of the picture. Some individuals see more of the picture than others, but there is not one man living that can be conscious at any one time of the whole, it is too much, the area in question is too vast for human consumption. However, every living man has the ability to see the whole in sections, and in time, will melt into a sea, the whole vast aspect of the universal picture.

We are working through a channeller with average vocabulary, average intelligence, a degree of sensitivity that can accommodate this. The words we use are words that average intelligent man can understand. But the conceptual forces behind it are those that you will have to reach to in your sensitive awareness, to be able to comprehend.

The velocity of your globe is being altered. This is changing the vibrational structure, and the difficulty in giving this information is that many of you are not aware of some of the facts of which we speak. Only in recent times has man become aware that his planet has a note, a vibration that he can tap into. Although that knowledge was there in past times, man is only just beginning to remember this and there are still some that deny it. So the difficulty in giving this information is to speak of things that are factual, that you are unaware of, let alone the difficulties of taking you beyond those facts of which you are not sure, to reach the conclusion of this particular evolution of man. We can only ask you to have open minds in the real sense of that word. We ask you to look beyond these words. We ask you to look to the concepts behind them. We ask you to do that, because only by doing that will you realise the heightened perception and truth that we speak.

Do not enforce your knowledge on others. Gently suggest, introduce it to them. Do not create a new religion - this is not information that can be structured and you will cause yourselves great pain if you try. Every man must find it within his own individual consciousness. Man must alone find his truth so he can, in turn, be a part of God in a conscious way. We leave you with that thought.

IV. UNSEEN ENERGIES

Partially the reason that the information coming through is so difficult for you is because it is about subjects that most of you have not consciously recognised. This is information about energies that the majority of your population deny exist. And even for those of you that have a sense of the energies, some doubt still in their own minds as to the validity of what they receive. But the energy pattern around your globe has changed, and the beings of Light that have nurtured your Earth on all levels from the physical to the higher mental planes are, as it were, changing shifts, so the work of the old energies is done. The old teachers that have done their job extremely well, are retiring and moving on into their own individual realms of evolution. Some of these energies will mix and mingle with the new for a while, but then dance on into their own higher Light forms. It has always been easier, particularly in the past, to personify these energies, to speak of hierarchies, to speak of group masters. This is something that the human mind can understand. In some cases it was not really accurate. However, it has served its purpose. The hierarchy, as some of you recognise the energy, is making its last bold attempt to heighten consciousness, stimulating objects, minds, thoughts and emotions in a way that in the past would have been considered interference. It is, in your terms, a kind of 'kill or cure', but the times in which you live, mean that there is this necessity for these actions. So you are seeing what you may consider to be the temperament of people being altered, being excited, sometimes running away from themselves. This is all part of the change.

Little physical manifestations of the energy are also permeating in the body and through into the conscious. Medical science will not be able to understand some of the illnesses that they are receiving in their surgeries. There will be growths that are not cancerous. There will be growths that exist of nothing and in analysis will be nothing, and yet they will register the growth. This is the energy frequency within the body changing and altering and creating little dumps within the physical body. They will die away, they will smooth out and they will not cause any major difficulties. We choose our words here carefully; there will be a spate of occurrences, of illnesses and malaise that will not be able to be explained. If you try to explain them, particularly in psychological terms, there will be no answers for you and will cause frustration and irritability. Your world is simply going through something of which it has never had any experience.

We have spoken about the karmic influence coming to an end, and it is so. The circle has turned full, and now you are moving to a higher spiral. This circle has been many thousands and millions of years. There is no record of human or any other information before that, not physically, not consciously, not mentally not at any level. So it has been that many souls are here today because they are very old souls, because they came in at the beginning. They can be the leaders to help those that do not understand at all what is going on. Most of you understand very little, so the necessity for these old souls is imperative.

We again try to put across the information in very simplistic terms because by the pure simpleness of it you can understand. But within the simple seeds that we sow of this information are seeds of enormous conceptual information. If you get a piece of this information, throw it up into your mind and let your mind escape, some of you will understand the higher conceptual information behind it. Words and speech has always been a lower frequency, and to describe these conceptual thoughts of the Universe into words is something that is simply not possible; we can at best only sow the seeds. So you will find that from time to time with the information, we will jump around a little. This is deliberate. By jumping around it creates an energy between the thoughts that accelerates and raises the conceptual understanding. Of course some will read and listen to these words and see the simplicity, and that is of value as well.

We are working with many channellers, not in the way that we are working with this one, but in different ways. We are working with those that are scientific, that are medical, that are inspirational and creative. There is a lot of interaction going on between the souls that can accommodate any of this information. Some of it will not be published, it will merely be for the enlightenment of that soul to transfer to the people within their vicinity and so forth. The upsurge of the religions, two thousand years ago approximately, was done in very small ways with very small thoughts, and look at the enormity of that influence.

Again we explain to you that the information can be received by the heightening of the consciousness and the opening up to the higher planes, and again we say, this is available for all. It is not a secret or occult situation. It is open to all those whose perspective can accommodate it.

We have spoken a little bit about the transmutation of the dimensions and the matrix of time that holds the dimensions of your Earth together. This means that the vibration of sound, the vibration of sight, of Light, is altering and moving. There are some people at the moment that are seeing twinkling Lights in their sights; they are seeing it in daylight, they are seeing it and they don't know what it is. These are the stimulations of the energy in the atmosphere, because in reality the energies are with you now, around you and running through you. It is stimulating, it is sparkling and can, for some, be seen by their physical eyes. This means that in a short space of time, some of you will be able to see differently, not quite in an x-ray fashion, but see through things. Far fetched as it sounds to you, when you are fully accomplished at working with the new energy fields, matter, as you understand it, will not be solid. So you will in time, all of you, be able to walk through solid structures.

This knowledge that is permeating could be disturbing. You are finding it difficult to hold on to your grounding and everyday lives. We emphasise again that the stillness within is imperative and you cannot go forward without it. By stillness we also mean you need quietness, you need a space. This is not always possible with the atmosphere of your noise pollution and so forth, but you can find a way. Five of your minutes a day will help you adjust and go with the changes. You are becoming free spirits, you are becoming free within the body, this is something that has also not occurred. This gives your body a completely different format and frame. It gives your body different work to do in different frequencies.

The energy that we have spoken about has been coming into your atmosphere for a little while now. It has come through first inspirationally, to the intuitive, down to the mental feelings, the emotional, and now is with you within the body. It will not be long now before your scientists have no choice but to admit to the energies beyond which they can see, because it will become crystal clear that something is moving and shifting, and it will take a very blind man indeed to deny it. Again, it is always difficult to give exact times, but the highest probability is that within two years science will change its mind to the unseen energies. Look out for the words "unseen energies". They will be spoken and they will be realised. They will find the scientific way of describing what you know to be the astral fields, and then they will find the fifth dimension of Light.

Remember we said to you in the previous communications that it was important for you to enjoy your life, to feel privileged by your physical being, the touch, the sense, the smells, the sights, the sounds. This information of enjoying your physical body is because the physical body, as you understood it, is no longer reacting in the same way. So it was - and we speak in past tense - an experience that you no longer can have. This does not in any way mean that you cannot enjoy your Earth, and indeed you must, but it does mean that you will not be able to enjoy it in the way that you have. Those of you that have been in contact with your bodies, particularly in the form of movement and dance, will realise this all too well.

There will be a breaking down of the immune system. The cells in your bodies are being altered, and these cells under a magnifying glass show no difference as yet, but they will in time. As these cells change and transmute, the body's defence mechanism is, to begin with, fighting this, just as it does when you pick up a virus, an illness or you cut yourself. This is sending the whole system of the body into disarray. We know it is difficult for you to understand that the body has a kind of mind, but it does. In these terms the body does not know what it is doing. This of course affects the immunity, and it means that there are people that cannot get better from illnesses, that the illnesses go over and over in a circular fashion throughout the weeks and months. In a small way you see it in your flu viruses, in a bigger way with other viruses. But the body is a beautifully accommodating instrument. You forget that your body has evolved and has only survived because of its evolution. When a new situation occurs it immediately works to accommodate it, and so it is doing beautifully now. There are, of course, those that are having difficulties physically at this time of change, but this will straighten. It is very important for you to recognise that your body itself is becoming stronger, not in terms of bigger muscles or strength of lifting and so forth, but the cell pattern is stronger. It has its different and strengthening DNA, as you call it. You will see.

In the past you have had great plagues. Your history knows about them. And in these plagues some survive and some die. It is not always possible to see why only some survive.

The affect of the ozone layer is that for some it will mean - and we emphasise in your words - a kind of radiation poisoning, but the opening up of the ozone layer is part of the change. We cannot say to you in all sincerity that it will not mean some casualties. Again, the word casualty is your terminology because there is no such thing as death. There will be some, again in your terms, suffering, but the body is very accommodating of its forces and always has been. So you will find that people will survive and will be strengthened by this. This does not mean that people must not take the natural precautions. One thing that is very important on a purely physical level is for people to protect the top of their heads, particularly those who have not much hair or have a particular skin structure. For it will burn and create cell deficiencies, creating the possibility of cancerous cells, but we do not alarm you and we want to assure you that all this is part of the tremendous journey that mankind is taking, and that is a positive journey into new states of being. The opening up of the ozone layer is allowing in cosmic rays, and these cosmic rays are essential for your development and growth. In time the ozone layer will repair itself. It will not be the same as it was before, but you will have a protective layer around your globe, but it will take a little time. Science can do nothing about this.

We have spoken a little of the increase of the growth of fundamentalism and nationalism when we spoke of pockets of negativity - pockets of energy being brought together like you sweep up the dust from your floors. As these are brought together, so it creates an irritated energy. This irritated energy manifests itself as what you describe as fundamentalism. It has to do this to work its way out. More so than our last communications, the negative forces around your globe have no choice but to be transmuted into Light. So you will see these pockets of fear that will be alarming, and some of them will cause destructive elements. But those of you that are in tune with your intuition and your inner note will not be anywhere near or a part of these situations. We spoke of a time when there was a lot of movement of people, of population. To a lesser degree this is again going to occur, within the next few months for you. You will see groups of people moving around the world, more so that usual. Again it will be to do with the adjustment of the frequencies. Those that are intuitively inclined will not be in the places of these dark pockets of energy. They cannot be, because their Light energy repels from the darkness. There will be, unfortunately, some breakdowns. You will see this in one or two of your political figures, you will see emotions, tears and distress. You must send your Light to all people of power now, because theirs is a most difficult path. The higher consciousness and the hierarchy have worked within the last hundred years or so to put into position the leaders that are right. Whatever your political or moralistic beliefs are, understand that largely every leader is the right leader, even when that means that the leader seems to be a destructive one. We are not saying that every leader is a Light being in tune with their spirituality. We are saying that every leader is the right leader for the time. For the leaders that have, in your words, a conscience, it is most difficult, and the strain and the stress that they have to deal with is harder and stronger than at any time in your history, including the leadership in the wars. So we ask you to send your Light and your strength to them because it is needed, and this Light will help their protection and their knowledge.

There is a vast opening up of the cerebral cortex occurring within the mind and within the brain. It is almost impossible for any of you to understand. It is simply an opening, an opening doorway, a shaft of Light, of energy forces that is entering. It is beyond any energy that you have ever received. This is happening en masse, and you will see evidence of the shake-up within a very short space of time, particularly in your leaders.

Never before has the thought and the meaning of 'live for the moment' had so much importance. Each moment is a stirring, atmospherically, on a cellular level and mental level. Each moment is a leap. Your time is altering and being charged. You are, in your terms, living in a different time. The speed of time, the ratio of time to the dimensions, is altering. So for each individual it is important for the stillness and it is important to go with the moment. Some of these concepts will take time and careful consideration, but we will leave you today with the very real knowledge that the population of the globe is being freed.

V. THE LADDER OF CONSCIOUSNESS

There is a clear space around your globe now, a clear point of entry for the new energies. Although they are not in fact new, they are energies that have come before. There has been an enormous turn around of the whole system, by that we mean both the solar system and the systems of other stars. There is a mammoth continuous turning around. 'Wheels within wheels' is the analogy we have used before, and it is an accurate one. Now, as the wheel turns for you, you accommodate and you open to these forces. Again, we say it would not be possible if work had not been done relentlessly for the opening of the doorways, the gateways in your outer atmosphere. The hierarchy and the masters within and without have helped and indeed, some people have helped. Sometimes it has been easier for individuals to help, not knowing how they are helping, but just helping as good souls.

So the forces have come down through the ladder from the highest consciousness, right down, and now you are experiencing that energy and the information down on the Earth level. This has come from the Gods down to the smallest creature, the smallest grain of Earth, and you are feeling the resonance of that energy. We cannot speak of a time when it began, because that would be beyond your comprehension, but it was many, many years ago that this whole process began. It really is a major turning point for yourselves, for you globe, for the system and for the cosmos.

As these wheels turn they touch off between each other, one wheel touching another wheel. It sparks off and loosens the barrier between one wheel and another, therefore communication of a type has occurred between the different systems. We speak in very simplistic terms here. When we speak of systems and wheels it is easier for you to comprehend. Information coming through the many layers of consciousness has come from different places and different energy forces, that in the past did not cross over your dimension. Some of these crossover points are not without or outside your own dimension, that is to say the crossover has happened within your dimension, but within a different space, and some of the crossover has indeed been between the dimensions, and that is a different thing again. So there is a cross pollination of energies now taking place on your planet. This is creating, even for unconscious man, dreams of a type that he has not known. Disturbance within the mind, even when he sees it not, is occurring.

As this energy force enters, because of the loosening of the connecting wheels, it is entering down the many layers of bodies and dimensions. Most of you reading this material are conversant with what you describe as the mental body and the emotional body. These energies disturb, and by disturb we do not mean negatively, but disturb nonetheless, right down to the physical. We will illuminate a little more on this, because this is how you will see the effect of the energy. For most people the first inkling of this is in what you define as the subconscious mind, and it comes in little flashes of intuition and awakening. It is what some of you refer to as 'the Light within', the flash point Light. At first it is ignored, and then gradually it comes more and more to the surface through your dreams and through your experiences. Very often, at this time, man has encountered what he considers to be incredible coincidences and synchronisations. Of course this is because of the opening doorways and the heightened magnetism occurring within that individual. Then it comes into the mental state, the mind and the thoughts. It is at this point that the individual knows something is happening. Incidentally this process has always occurred when there has been spiritual awakening. We speak of it specifically at this time of change because it is in this time when it is not only happening to a few, but it is happening, to some degree, to every living being. So it is a mass rise of consciousness that in the past has only happened to a few.

We have already spoken in previous works of the difference in magnetism when the acknowledgement and the awakening occurs on many levels. This is when change of frequency occurs. When the mental body is stimulated, it is obvious to you how that affects - through the magnetism of your changing thought patterns. The patterns of your thoughts matter a great deal to the outside world and to how the outside world reacts to you. So automatically changes occur in the life of that man. Then it comes down through the process of the mind. Then it usually hits the emotional body. This, of course, is tied up with the astral plane, or indeed has been in the past. This emotional body is both within the body and without, in what you refer to as your 'shadow self'. We spoke very briefly in the last publication that all emotions are negative. This may be disturbing information for you, but on the mixture of emotion and love, we speak singularly of emotions. They are largely destructive because tied to the emotional state is the link to the shadow self - the astral body. All of you that are truly awakened have to dissolve the remnant of your astral bodies. This can occur automatically, but in some cases you need help. There are healers working with this now. Some of these, incidentally, do not know what they are doing when they clear what they consider to be a shadow self. On some occasions they see it as an entity, when in reality they are clearing some of the negative energy. This is an important part of the work of clearing and awakening. It is setting free the body.

The heart energy is the energy that links above and below, and becomes, at this point, very much the base of the energy in which you are to work. The emotions are something that you needed as part of your karmic journey. Through the emotions you have learned tremendously. You have grown and you have prospered, and for that reason some of you feel that emotions must be kept. But if you can separate the emotions from the true aspects of love and the true aspect of the Earth force, you will see that it is an insular force created by yourselves for yourselves. When you were evolving, the higher consciousness, even what you call the master rays, knew not what would occur, and the emotions were, we will say, a surprise element. They are now diminishing, dissolving, and it is right that they do so to enable you to work on the higher planes. As the ladder comes down it touches the physical body through the etheric body, that bringer of Light that is your blueprint for the physical being.

It is to the physical changes that some of you are most concerned. Even at this point the rate of your physical change, the rate of your cellular and molecular structure, is altering, is not fixed. That is to say there is no complete knowledge of what will occur at what time span. But we say that the nucleus of your cells which is, if you like, the seed of your new germination, is being magnetised and de-magnetised. It is at this point when some of you awaken to the conscious reality of why you are here and what you have done before. Because within that disturbance and awakening is the information of past times and energy fields. This nucleus disturbance is similar to what would happen if you have a genetic engineer working on your cells to create what you would consider to be a superman. The superman is not one of physical strength, but is one of a Light being. We touched on this in the work of 'The Cosmic Dance'. This information is difficult because immediately we speak of super or above, it creates for some people the disability to look at the real possibilities and potential of man. We urge you, once again, to look at these possibilities, because until you look at them you cannot acknowledge them.

At the moment, through the dream world, through the astral plane, we are helping to align the dreams of man to superman. We need the dreams of man, the wishes and the true desires which are unemotional, the true note of desire, to project and to push forward. Without that there will be none of you moving into the new worlds. So, dear children of Light, we need your dreams, because through that state of the astral enlightenment and the astral yearning, Light will dissipate the darkness, dissolving the barrier that holds you from the reality of becoming the superman. Within these words there is much.

When we speak of working with people through their dreams, we are not in any way talking of what you would call the psychological aspect. We speak of the desire for perfection, the desire for coming home, and with that energy, with that need and the reaching, mankind will create the ability to reach the top of the mountain. A man only climbs to the top of the mountain when he knows he can, when he knows it is possible and he wants to. If he tries to climb the mountain without that note of desire, it is most likely that he will not achieve his aim. So what we are asking is that you inspire people to dream, both in the waking state and in the night state, of the possibilities and potentials of man.

The genetic seed we speak of is not connected to the deep desire, the potential of superman. The genetic seed is a disturbance within the physical cellular structure which opens up the pod of the seed sown many aeons ago for the potential of the new race of man. Now, this is difficult because what we are not saying is that it was a necessity to have this genetic possibility always within. Because we were using the analogy, that what is occurring is, as it

53

were, as if you had genetic engineers working with yourselves. There is a disturbance of energy within the cell, and we use again a picture, an analogy. It is creating a pod to open, for a seed that is different to the seed of your genetic imprinting at the moment. So it will create for you a disturbance of your own cells, and when you give birth to the new race they will be born instantaneously with the new genes. This, of course, is not always an actual statement of fact. We speak in terms that you will understand, but when we say there is a new genetic coding coming to the babies of the future, this is an actual concrete statement; that the babies of the future will indeed have a new imprint within.

Largely speaking, if people need help they will come to the souls that can help them. None of you need feel you have to make it your mission to walk out and heal. Just be available. This is all you can do, and it is all we will ask you to do, creating again, as we have often said to you, a perfected Light within yourselves. Creating a clear energy force is an extremely strong healing force. If another soul is needy and is going through a process of awakening that you may be able to help, then that soul will come into your pathway without you doing anything specifically. However, if any of you truly have the desire - and we speak deliberately with the word desire - to set up this or that centre or healing place or aspect of healing or intuitive work, you have a duty to follow that path. But unless you have that real intuitive knowing, knowing from the soul level that is what you must do, we ask you just to be still and to polish your own stone of healing, creating an energy field of Light that works as an automatic healer for all those around you. It is hard sometimes, knowing that perhaps you could help, not to feel compelled to knock on doors. But you need not knock on any doors, because those that need your help will knock on your door. You can be assured of that! As you melt into the awakening process, you can be certain that your desires are what is needed, because the desire that used to be on a lower level of want and selfishness becomes attuned to the soul note. As it becomes attuned to the soul note, that desire is a desire of purity and it is a right desire, and the desire becomes right motive.

We have to, as it were, assimilate our thoughts to push down into the channeller this simple process of words that can have some meaning when we speak of concepts beyond the mind.

The DNA structure within the body is changing. This is carrying on slightly from what we spoke of before. This DNA is the building blocks of your human structure, and as the energy coming from above your planet alters, it is automatically altering the building blocks of your lives. You have never considered the possibility of what you describe as the DNA being connected to the higher frequencies, the higher consciousness, if you like to God. But it was and always has been. If you take on board that reality, you will see also how the changes are occurring within your bodies now. Within all these words that we give you is the potential for opening up the mind.

Some of you would deny the possibility of linking with the mind of God, and in the past, for the vast majority of you, it was not a possibility. Now you have this attunement where you can link to the mind of God into the source of Light, and this means a disturbance of the mental planes. As it becomes attuned to that mental frequency we cannot say that you will all have the knowledge of the higher consciousness of God. What we can say is that you will have access to part of the library, so to speak, the inner door of the mind. This is only possible because of the breakdown of the wheels within wheels that we spoke of earlier, only possible because of the vast acceleration in the evolution of man. It will never be possible for any individual being to know the whole process of the workings of all the energy forces in the whole cosmos. Whilst he considers himself separate from God, he cannot see the whole. But you are becoming less individual at this point of your evolution. You are unifying in your energy frequency, and as you unify, you get a sense of detachment from the Earth, detachment from the individual personality. As you feel that detachment you also feel part of a whole union. This union is not just a union of your fellow man, it is a union of other beings of Light. The union to the other beings of Light to which some of you are already referring to as 'Light beings' is a wonder for you. It is, in very common terms, like meeting a very special person that you want to embrace because you have caught sight of their energy. You now have the opportunity and the realisation that they are close, and they are becoming part of you. These Light beings, in consciousness terms, are with you now. Their consciousness is here, and it melts and accumulates Light with you as your consciousness accelerates and lifts in Light. This is creating a vibrationary note that in turn helps the whole of your globe and helps the whole of theirs. Now we speak of them coming as a consciousness. This is a difficult acknowledgement, because their consciousness can move in and out of their physical. There is no real purpose in us telling you the way that their bodies are, because you have no framework by which to map. We speak of this now because many of you are recognising this union of Light beings. Some of you are considering this to be the most important aspect of the change. It is not, dear children. It is almost, you could say, a by-product of it. It is as though the telephone wires are open between your world and theirs. This will not always be the case, because as the wheel turns further again you will go about your own separate business, but the joy of being able to communicate and contact these other Lights at this time is rather pleasant.

Some of you have been aware of a darker force coming that is not to do with the astral. We assure and confirm to you that this dark force could not exist in your world and has now left it and was not like any dark force of the astral.

So enjoy your communion with the Light beings. They dance on your laughter; they do not know emotions. When you touch them, you are getting a glimpse of the possibility of your future. And it is a possibility because as the DNA is opening up, unbeknown to you, you are creating, by your desires, the future. You are shaping the new superman. In catching a glimpse of these Light bodies you have an inkling, and you put some shape to these possibilities. It is, dear children, as though you are given the clay to make the form. So you recognise now the very real need to have, not only positive thought, but positive desires and minds. You give shape to the new man. It is not, at this point, a haphazard automatic process. It is done out of your higher note of desire. Now do not concern yourself that those men that are still low in energy and are evil in intent will have their say in the shape of the new man. It is only registered in the heightened desire and, as we spoke of, that heightened desire is complete with and is unified in the Light of the highest note of the soul. In this is a huge bit of information for you, and to some it is, in your words, far fetched and beyond reality.

All through your history, your acceleration of evolution has only been possible by those of you that have been inspired and have allowed themselves to be inspired by the higher forces that create a magnetic energy that lifts the whole. There are always leaders, and these leaders are leaders in inspiration attunement to Light. They do not always know it, and are in your terms the genii. Unbeknown to all of you, some of these genii were not made public, they had, within their hearts and souls, the truest desire, the altruistic desire of man.

Lift up your hearts, lift up your minds to the possibility of Light, all of you now. Once you lift your mind to the possibility of these extraordinary concepts, you are making them and you are moulding them with your clay. The material of the clay is in the Light, is in the magnetism, is in the thoughts, is in the desire, is in the mind of God that you now touch. We leave this for you to assimilate.

To break up the old - and a break up is a necessity as most of you now realise - you have to break up the hard and fast structures that you have built for yourself. The financial world is such a structure, and it has been imperative for the system that you have worked by, to change. It is centring people's thoughts in a way that was not possible with any other situation. Man has reached a point in his evolution when it is just not possible for him to go back to the grovelling cavemen of the past, because the energy is too fine. So if you take away the material there is no danger of that. You have concerned yourself that man would become grabbing, violent and aggressive. The vast majority cannot, because the consciousness is too heightened. This means, by a very real lack of the material, they are being instantaneously snapped back into what they have, and what they have is within themselves. What they have is their souls, is the energy and the acknowledgement that all is well, that they cannot be harmed, and even if they should die physically, they are all right. We do not speak specifically on these kinds of subjects because it will create for some of you, unfortunately, the leaning on of a 'prophet' type information, and this is not our intention in these communications. However, it would be foolish to say to you that this recession you are now encountering will improve or will die away. It will not. It is changing the whole process of your material lives. There was a possibility

58

of this happening after the first World War. We spoke earlier in the session of the energy coming down the ladder. Well, you can clearly see that it has hit the material by what is transpiring now. Man has accelerated so much. There are some leader men that will know what to do for their races. All negative pockets of energy need to be drawn up, to dispel into Light. These have gathered in many institutions, including the financial, so this needs to be broken down. There will not be one man that lives purely for the material that will not be affected at this time - and it must be so. This is not a punishment, this is not any sort of evil intent or desire from the higher forces, it is a necessary process of breaking down to build up.

We will leave you now, but before we do we urge you to give some consideration to the aspect of the desire of the soul and the inspiration into the making of the new man. You have helpers on the higher planes to do this. Think on this.

VI. STARLIGHT ENERGIES

There is feeling of urgency in the atmosphere, many people feeling they have to do things in a hurry, a sense of having to have a purpose, to be somewhere, to do something quickly. They are not wrong because there is not much time in your world for making the changeover. We can say to you now that those that have not put their toes on the pathway of Light will not make this journey. This does not mean that those that have no conscious knowledge of the journey will not make the transition, it is those that have not started on the pathway of Light. This could have occurred, of course, in previous lives. However, those that are dead to this knowledge, those that do not see or experience, or have experienced in one way or another, their time is running out. We have to say to you now, you workers on the healing pathway, that those that have not made this commitment of spirit, you cannot help. This is a difficult concept. This is something that you do not readily understand, but you will see for yourself. However, if there has been a glimmer of Light within an individual, they still have a chance to move over and embrace the new source of energy that is here now. It is no longer, as we have said, a case of if and when. It is here now.

In physical terms the possibility of this energy coming through is to do with your physical atmosphere changing. You have long had the influences of sunlight upon your globe. The sun has been your creator on a physical level. Now, as the atmosphere changes and the Earth is ready to make its tilt, it is accommodating the energy from other stars. And as the transition and movement occurs, those other stars that are in themselves suns begin to radiate their energy, their frequency and their Light, creating a disturbance and a regeneration of spirit within all those that live on your planet, the plant life and the animal life included. The energy of those that are dead to the transition of humanness into spiritual awakening, will, like the leaf being torn off the tree, implode. This is an automatic procedure. To speak of mass death anyway at this time would be counter-productive. We are saying that there is a mass movement. The leaves are being blown by the wind, and they will find, by their magnetism and energy, their place. This is how it will be.

Those of you who are sensitive, who are open and intuitive, are the ones that have felt this first. Those intuitive ones are now getting connections and communications from, in your words, the White Brotherhood of other places. You have long worked with what you define as the Hierarchy. You now are beginning to work with the communion of White Brotherhood that comes from other places, other star systems, in your words. This sounds to you like science fiction. It is not. It is a reality. There has always been a communion of work done between all those Light workers within the cosmos. Some of these Light workers do not know of the existence of other Light workers, and yet they work in unison of the heightened consciousness, just as you as individuals work together on the higher level even when you do not know.

Unknowingly, you are working with other consciousnesses, other beings, as your energy has radiated like the ripple, moving, touching and communing with other consciousness. There is in operation always a group consciousness, and it is not necessarily the case that either the individual group or the individual soul knows on a conscious level that they are doing this. So now the group, in the old term of guides, of this planet are in conscious communication with other guides, other brotherhoods from other places, as the energy of their star, their sun comes to impress and impregnate your planet. This is, in your terms, an alien force, but alien only means different, it is not in any way evil. This means that particularly the sensitive ones are coming in contact with guides and hierarchy masters, and also cosmic beings that have different information. By different we do not mean opposing. This information is not alien to your White Brotherhood, to your hierarchy, but it has a slightly different perspective.

Your Akashic records on the astral planes are dissolving. They are being absorbed by those that can absorb them. So some sensitive people can readily lock in, and indeed, it is as though they have that whole computer of the Akashic records within their brains. As that diminishes from the astral plane you can connect, with a similarity of records you could describe as the Akashic records of other places, through these White Brotherhoods. Some of you already recognise and acknowledge the description of 'White' as different from 'Light'. There are Light beings, but they are often depicted and sensed as White Beings of Light. You know this energy, some of you, and you are beginning to work with it, and it is important that you do so, because this energy frequency from the White Brotherhood - we will describe them as such - is an important transitional energy.

It is important to construct and reconstruct the frequency within the body, and some of the healers will use it, some knowing and some unknowing. It is electrical currents of Light, and it pours through the body changing the structure. We spoke of the heightened consciousness, the higher consciousness connecting the DNA. Now as you have moved and your planet has shifted its atmosphere you are ready and are beginning to contact this new force. You are able, through the heightened consciousness of the individual and of the group, to contact and to stimulate the cells within the body, connecting these and altering it. Do not be alarmed at this. Speaking of altering the DNA to some will sound as though you will transmute into some mutant beings! That, of course, is not the case. You are transmuting though. You are transmuting into spirituality within the body, a rare occurrence in the cosmos - and again we speak of the true privilege and experience at this time that you are encountering.

Can you not see, can you not realise the tremendous possibilities of the combination the spiritual and the physical bring? Yet some of you do not consider this to be a possibility. Spirit has been one dimension and physical another. And although you recognise the soul and the spirit within, you still consider them to be two different dimensional frequencies. Yet the awakening and the growing of the spiritual essence through the etheric body into the physical is now occurring. It is, to you, as though positive and negative are clashing and combining. Matter and anti-matter becomes one. Those of you scientifically based, wipe out that possibility. It is indeed rare, but in this cosmic changeover it is happening. Some of you will keep your body in this transition. This transition has happened before, but the difference being then was that it was not possible for the body, the physical, to be kept.

This of course means a change of structure of your cells, to accommodate the spiritual mass within through the etheric being. So some of you, and more as time goes on, are experiencing the possibility of shaping, through a kind of healing work, into the etheric body, making the new blueprint for the physical. It is hard work for you to recognise and for you to open your minds to that possibility. As we give you this information we know, readily, that some of you will run away from it. So be it. Even some of those that run from this information will accommodate new forces within. The conscious mind is not so important as you think it to be, and it often runs very far behind in the race for growth. It is not far removed from the heavy mass of the physical, and therefore it has always been something

63

that, in the evolutionary process, has acted like a lead weight on your heels. It does not matter whether you can consciously accommodate these facts, they are working anyway.

The White Beings of Light want to introduce themselves and are very close. They are new to the dense physical matter that you occupy. They look upon you with the very wonder that you look upon them, because their experience is different to yours and their teachings, their knowledge is different in the worlds that they have accommodated in the past. We put into your mind the phrase 'a banana republic'. You will notice growth.

As we have spoken before, the information coming through this particular channeller is one to enhance and project, like seeds. Others will take on the concepts and go further. It is not the work of this information to do that, it is to sow seeds of enlightenment and higher conceptual thoughts, so you can read these words as a matter of fact. But they are also projections of higher conceptual thoughts, for those that can take them on board. It is the difference of the ordinary arithmetic to the higher mathematics. Think on that. Sowing seeds in the conscious mind allows, through the subconscious, the super conscious, to accelerate and to grow and to transform those seeds into higher thoughts and higher possibilities. That is what we give to you today.

You wish to know about centres, the work of the inner movement of energy, and what the wheels of the sharper points are doing. We cannot tell you exactly what will transpire, because the changing of the DNA structures will bring different points of energy within the body that will change as the adjustment is being made. The hub, the core point of energy that you experience now may not be the same, even in as short a period as a few weeks, a few months. Basically the sharp core points of the body still exist. They are like stations. They are entry points, but like stations without trains running through them, the energy is stilled, but they are still available for the energy, should that be the necessity. Of course this is very much to do with the individual. In general terms the lower energies, the lower frequency centres are becoming like stations without trains. You will understand this. Surprisingly, from time to time there will be a force within those lower centres, but it is unlikely to last. A period of adjustment of frequencies is transpiring as the new energy is here within the body. It takes a large degree of trust and heightened intuition to know as healers what to do, and in some cases healing will not be possible in the ordinary way. A form of healing is occurring through the frequency rate, and many other people are working with this knowledge.

The frequency rate is being heightened first through the auric emanation and then the etheric body, creating this blueprint for the new structure within. This will be done through the heightened energy of the healers, their connection with the new forces of Light working with their energies, the White Beings of Light, injecting within the etheric, subtle energies, electrical frequencies that are required. It is not necessary for the healers to know exactly what is occurring. They will do it automatically, if their frequency is right. You are very correct when you speak of the necessity for trust and for faith in your own energetic force within. We give you words, but we do not want you to label things. We, through this communication, are doing what we can to break down preconceived ideas and structures. We do not want to create new ones. However, we have to speak in your language, and your language is not so good for giving conceptual ideas without structures. We do not want to set about creating a new structure of healing for instance, because every individual must know and realise for themselves what is correct, and what is correct one day may not be correct the next, as the frequencies themselves adjust and move. So the importance of the information done in the previous two publications has been important to manifest within the individual their sense of their own Light, their own energy, and to realise that even when working in groups, both big and small, they must recognise and listen to their own way of working, their own truth.

If you could see your cells, they are bubbling. They are bubbling like molecules in boiling water. They are transmuting. This should not frighten you. Ice transmutes to water which transmutes to gas. It is the same substance. It is the same essence. You are transmuting physically to spiritual. It certainly should not be a fear, and should there be any worry or fear left within, we urge you to eradicate it completely.

We understand and we acknowledge that you want to get things right. Within this time of hurried change you must acknowledge that working with your inner Light you cannot get anything wrong. You must know it to be so, because in that knowledge all doubt diminishes and dissolves.

This information we are giving you, will be published within this year, because by the end of this year, your year 1992 and by the beginning of 1993, you will be in no doubt of the enormity of the movement. We largely do not like to give you specific times, but we can tell you that a major shift will have occurred by that time. It is surprising to you that there will still be some beings who will not know this to be so.

++++++++

The resurrection that the church speaks of is the movement into spirit, as you define it. But the difference that is occurring here is that this transmutation is accommodating the physical body as well as the spiritual, so it is not moving from the physical body, allowing that to die, to move on to the spiritual. It is happening in both ways - a transformation. It is a resurrection, but it is not leaving the physical body behind.

Many beings of Light have worked towards this transmutation without even knowing the end results. Indeed, there is not an end as you well know. Each being of Light, each master being has helped in his or her capacity to accommodate the enlargement of the spiritual concepts, the enlargement of the spirit within, the enlargement of the consciousness, bringing energy down through and enlarging from within the body in all sorts of ways. As you know, in the past some of these ways have been very structured. But it is not possible for very structured things to work, and anyone who is working in a rigid way will have to change. Groups and past teachings, left behind, will have to change their ways. It is all one giant movement - a dance of energy shifting and flowing in the cosmic Light.

There was a lot of work done by souls to link with the core of your Earth. Within the core of your Earth are enormous eruptive energies. They are, in your terms, volcanic. They are power, they are force, and those volcanic energies will stay. They, of course, are shifting also in the change, but they move within their own circular body. This energy force can be locked into, not necessarily through the base chakra now, but still can be locked into through the core of the Earth. You can use that imagery of the ball of volcanic mass, a vibrational frequency within your Earth, to connect with that ball which will, in a sense, anchor the Earth energy within. To speak of grounding is not quite accurate. We can speak of anchoring the Earth's energy within the body, which is different. You can do this, not necessarily through any particular chakra, although you can do that as well. Your brow chakra and your crown chakra can be linked to this core energy, because the volcanic eruptive force there is similar to the volcanic force within the brow centre. There is a kind of magnetism that always has run between them. Again, we are wary about giving you individual structures and ways of working. Again, we say trust your own judgement, but it would be easier for you to talk in terms of anchoring rather than grounding. The effect is similar. Just think of the core, the ball of red, yellow, vibrant force within your globe. Although this ball is moving, the energy will stay the same.

In truth there will come a point when the chakras do not vibrate. We describe them as stations without the trains, and eventually, if there is no movement within these stations, they will become as 'ghost stations', they will become inoperative. The land will grow up around them. There will be a growth of weeds and flowers, as it were, covering up what was there before, and that is occurring very gradually. Of course, in this sense we are talking about a length of time, and there is no chance at the moment of your centres becoming what we describe as ghost stations, but in time they will. The lower ones first will go, as we have already described. There will be entry points through the outside of the body through, again we loosely say, the etheric - loosely because that it not exactly as you imagine it to be. You will be able to accommodate energy through entry points coming in similar places as before. But we do not want you to rekindle old ideas, although some of them will melt into the new concepts. It is difficult for you. Those that are working in this field will know what to do, and even those that work within the field that don't know what they were doing exactly in a conscious way will be doing right anyway. We have not spoken about the changes occurring within the aura because we imagined that it would be obvious to you that as the inner energies change, the auric emanation changes. There will be points of entry into the chakras to help the transformation of energies. Be free in your ideas, we ask you again not to recreate different structures of approach, for if you do you will find that the new structures will not last.

You may be feeling that some information you receive at this time is contradictory. As information comes down into the physical mind it is often distorted. Do you know, for instance, that what you dream and what you remember of your dreams is not what you actually dreamt or actually experienced? Think on that, and you will begin to see how the disturbance of the human mind and the atmospheres that the information comes down into distorts. You will find however, that within every piece of the contradictory information there is within it some seed point, some point of reality. We have spoken about the different ways of perception from the point on which you stand, and this is so with the mind and the way the mind distorts. However, in a very positive way now, you have the ability to see things in a much clearer state. You must take the information coming to you. You must let it run over you. You must listen and acknowledge the seed of this information. Take that to your hearts and walk on.

Communication with the White Brotherhood, having been made, cannot so readily be broken. It is as though you have a transmitter pointed at the right place, and having found it once you can come back to it should you need to. So there is no chance of this diminishing. These White Brotherhood beings are with you now, particularly through your transition, and you can and you will link to them. They are friendly beings of Light.

You will know the expanse of the starlight coming is a radiant beauty for you. You have, through your long history, sometimes worshipped your sun, and your sun is a small star in the cosmic scheme of things. Now you have the opportunity of the radiance of Light from other star places. Enjoy that Light. Experience it. It does not matter where it comes in the cosmic chart. You can give it a name. You need not give it a name. It is up to you.

VII. CHARIOT OF LIGHT

We are as you are, everything and nothing. Already man is realising the necessity to break from the desire of requiring personality in his connection with the spiritual or higher realms. The need for names is dissipating. It is not necessary to have a name. It is not necessary to have a personality and likeness. You speak of man being in God's likeness. It is so. All is one. You sometimes require descriptions of energies in your words, and you use names, and that is alright as long as it is just a word to confirm the energy that you work with. Do not be tempted down the road again of personality. We say to you judge not by the need for the personality, only by the resonance of the truth within.

Your energy is shifting both personally and globally again. Your energy is like a sea at the moment, creating waves, and the waves are coming to new spots, new parts, they are covering new ground. Do not be alarmed at anything that you encounter.

Some of you will unfortunately come across a type of madness both in the animal kingdom and in the human. Some of this madness will be temporary - it is the adjustment energies. Sometimes it will be permanent. It has the look of deep distress, but it is a flux of energy that may not create pain within the individual. Largely, if the individual clears and has worked with themselves, they will not encounter this, but it will be distressing to see this kind of side-effect to the energy forces.

You have a key point within the back of your head, a key point of energy and sensibility. Now you can, if you want, call it an eye, because it does see, it does perceive.

This is, as it were, the eye of the needle through which you must go to encounter the cosmic fields of Light.

Starlight is coming now on your Earth. You are in a period now where your Earth is very open. It is like new skin growing, but it is delicate. It is as though the Earth has been dug ready for new growth and it is sensitive, like a sponge. It is porous, sometimes holding energetic forces. The Earth has been in this state now for almost two of your years, as it is being 'dug over', ready for new growth. The energetic flow through the body of energy is one that is affecting the whole system, particularly the nervous system and canals within the body that help the flow of blood. There are surges of energy, and as these surges of energy come, it creates slight blockages of these canals within the body, but again these will clear. So you will be getting many different types of disturbance, but it will have the same cause, and the body will adjust.

We have spoken a little on the blueprints, the etheric body that is also being altered at this time. There are Angels of Light working with the etheric body, and have been doing so for some time. We speak of Angels as a term, as a word, as a description. These are not like Angels of the astral planes that you know. They are from other places, places that readily know the type of movement and change you are going through. These angelic beings are moving and healing, correcting, enhancing, strengthening the etheric body from without, so that when the individual has cleared themselves to a degree, they can breath in the clarity of the etheric body, mirroring a perfected Light. Some of you have been a little bit aware of what has been occurring, but not many, because it is in a sense so far removed from your field of physical energy that you do not perceive it. When you sense it, it has come down through all the levels to the physical, and this also is a reason for some disturbance. For change and transmutation, the etheric blueprint needs to be cleansed and altered. It is, for you, like angelic surgeons or angelic sculptors moulding the new force.

As you have worked helpfully and rightfully on yourselves, delving within, cleansing and clearing, moving forward, being spiritually attuned, you are able to accommodate the new force that works down through the etheric. This cannot be done without some degree of clarity within. If the etheric body was breathed in by someone that was unclear and unclean, this would create a 'billow-out' situation that could make for death. But the etheric beings know what they are doing. We have spoken about the need to let go of the psychological aspects and the delving that goes on within those areas. Because when you get your clear being through the etheric body, if you, for whatever reason, think and recreate these psychological needs, you are undoing all that has been done by yourselves and by these beings of Light. You would have, quite literally, wasted their time! Of course we never choose to do away with the freedom of choice. You must and will always have this. But we ask you again not to create illnesses and disturbances from the past. Your planet has moved over the threshold. It is working within a new energy field. It cannot and will not accommodate the old forces now and if you try to recreate them you will be in a great deal of difficulty and can, in extreme circumstances, create immediate death.

You know that in your evolution your bodies have changed. You know that man evolved from getting around on four legs to two. You know that horses had different fingers and limbs. You know these things. What is difficult for you is to realise the rate of evolution at this time, because a transmutation is happening in such a short space.

73

We would speak on time because there is much to say, but the degree of knowledge we can impart depends on the understanding of man at this time, and even the brightest of you will not be able to hold the whole concept of time. Some of you speak rightly about time having no place - there is no time to speak of. That is so. So you are curious as to why it should be that you have to work within time, but it is your time zone in its own bubble in the cosmos. Your planet is a planet in between, it is in between positive and negative energy. Your whole system, most of the stars you see in your sky, are part of this interim or interval between positive and negative Universes. This band, between the positive and negative, is breaking up, as you move very slowly from the band of the positive into the band of the negative. This movement has in fact been occurring for many thousands of years. You speak rightly of 'as above so below'. What you as human beings have played out is the dance of the cosmos in human form. As your whole system shifts, it is coming out of this middle band, into the negative side of the cosmos. So again we speak of different reasons for the enormity of the changes occurring.

The whole cosmos moves - there is not a static state. In simplistic terms there is a movement shift of the part of the cosmos that is positive material. A part that is negative, seems to you like two sides of a coin. It seems to you like a ball, but you cannot possibly even imagine the enormous cosmos of which you are part.

Planets and stars do not stay the same. Many of you feel you would like to return, but you cannot return, dear children of Light, to a planet that no longer exists as you knew it. Planets change and move. Energies shift. You are living in extraordinary times.

Because your planet has been in this interval space for sometime, it has adjusted itself without destroying itself. Your planet is very adaptable and will survive the change. In simple terms, time is a band that holds the forces, a belt, an interval place. It is correct to say that you are moving out of time. Because of your planet's experience, because of the energy fields within it, it is being looked after very carefully by energies above and beyond yours, because they recognise the value of this globe that can be a planet of healing, a 'holiday home' for the cosmos. In all this we ask you not to lose your sense of proportion and your sense of humour which differentiates you from the sluggish living beings on the lower level of evolutionary planes. It is that humour that is the best of the energy from your solar plexus that you take with you. As we speak of losing energy, of transmutation of energy, we say you will take the best of the old, bringing it and expanding into the best of the new.

You have been concerned about your atmosphere. You have been concerned about your ozone layer, that without it leaves your planet naked and vulnerable. Do not be concerned. It is all part of the movement, and in one sense is part of the plan. This is allowing heightened frequencies to enter. This is allowing freedom.

It seems strange to think that the Earth en masse is going to realise its need for protection and begin to love itself just as the planet changes. Whatever you do to help your planet it will not stop the changes we speak of, because they are not connected. The changes are only a mirror, they are helping to accommodate the new forces, but they are not necessary. They are, if you like, a side-effect. It may seem very negative to say to you that whatever you do will not change the situation very much at this point. But it is important for the vast majority of humanity to realise that this planet, this Earth is an entity. It does exists and it does need consideration. The way you have treated your planet most of you would not treat a dog. If you see your Earth as a being in its own right you must give it consideration. You must not hurt it. You must take care of it. This is good, because even though some souls will not incarnate any more on this planet, as it transmutes, the souls that incarnate to other globes that are similarly karmic in nature, will bring to it the realisation of working with the Earth, the ground. Nothing is ever wasted, no knowledge, neither small or big, ever goes to waste. No experience is ever without its purpose in the long term evolutionary process.

Many of you are aware of what you describe as Atlantis, that turning point in your Earth's history. Many of you speak of Atlantis as being the idyllic place, the place of knowledge, the place of vast education, using the energies that you are now beginning to use once more. But has it not occurred to you that Atlantis sank? Has it not occurred to you that they were not perfected, that they had their evolution to go through also? There are many of you incarnate now that have the memory and the experience of what went wrong. In many ways the Atlantean era was more to do with the fight of good and evil than this one, because in reality this change has nothing to do with good and evil, because that war has been fought and won. It may sound callous or unrealistic to speak of the automatic process of change, but it is so. The movement is in operation and cannot be stopped. However, it is to your benefit and to our joy that some of you have reached to the stars in your hearts and souls, that you have bothered to evolve and work to enlarge yourself as spiritual beings. If you hadn't, this planet would make its journey without humanity, but to your credit, dear children of Light, you have moved enough so that some of you will transmute with it. But again we say, even those that don't are not wasted. In their experience they will learn, will move, will grow.

We are picking up your concern about food, and over the next few weeks there will be a lot of conversation about certain foods and what is done to the foods. It is ironic, in a way, that as you become more sensitised on one level to what you take in, on another level you become capable of consuming almost anything. We speak very simply when we say "eat what you feel you want to eat". Some of you have gone through a cleansing process with food, you call it detoxification. That is fine, but do not get fanatical with what you eat. Know that what you really want is what is really right for your body. Some of you reaching heightened consciousness level will, quite literally, be able to eat anything. Think on that! Food, even the basic food substance, works by the same influences of magnetism. You will not eat anything you are not meant to eat. You only attract what you need. We speak in this instance of need as being the attractive force of energy.

Your foodstuffs though, are going through their own transmutation. The plant life, the animal life is also changing and of course that has an effect in itself which is all part of the process. There will be some things that people have been able to eat for many hundreds of years that they will not be able to eat now. There will be much spoken about the change in taste. The change of taste is tied up with transmutation of the food.

Some of you are locked in to the heightened consciousness. This is a universal consciousness. Trust it. Test it. Experiment with it. You will see. You will know. You will have access to the huge library of knowledge of the whole. This is not what you define as the collective unconscious, this is Universal consciousness. In these channellings we try not to get involved with too much detail. Remember, dear children of Light, you are free entities. Now, more than ever, you have the reigns of destiny in your hands. Take that chariot of Light and fly.

VIII. COMMUNION OF THE WHITE BROTHERHOOD

You all are entering now the hardest point of deliverance. It is limbo land for you now. In many ways those that get on with their mundane chores and tasks are the fortunate ones. They see nothing of what is occurring, and for those of you that know the promised acceleration of Light, it is not so difficult either. But for those ones that are teetering on the brink of deliverance, this is the difficult time.

There are many, of course, that cannot now move, and in reality their energy will gradually dissipate, like a leaf pulled from the tree. The heightened frequency state in your atmosphere is now creating disturbances. This disturbance is part of the cleansing, and healers and teachers are needed to impress upon those people that you are dealing with. The medicine is unpleasant, but the medicine is good.

Your planet's energy is disturbed and is fighting against itself. Its magnetic force is pushing and pulling now. It is fluctuating. It will be revealed that there are difficulties. Scientists are going to be looking above and below for the answer to this, because there is, quite literally, a magnetic disturbance. This is pulling. As the Earth tries to adjust back the consciousness is resisting in its change, just as your bodies are resisting. It is a bodily functioning that is occurring now. It is a bodily reaction, and although you are seeing some disturbance emotionally, it is actually in response to the bodily change. The energy has come down, right down to the physical now. So the disturbances are on that level. This will mean many things to many people. Those that are sensitive and are going to make the change will react in some way. Those that won't make the change may not react, as their energy force is cut off from the main stem. So it is not so easy, in logical scientific terms, to see this in reality, because there will be so many different reactions.

Although the information we have to give you now is startling, it is not to give you a reaction, and indeed those that read the words have largely lifted themselves beyond that fear energy. It is not possible to get a remedy for this, other than being intuitive about it. There is no remedy for what is occurring. There are beings helping on the finer levels to accommodate the new frequency within the etheric, as well as other finer frequencies, so that this new etheric field rains down upon the Earth, stilling and calming.

There are many beings of Light above and below working constantly to help the change. There are even human beings working to help, unknowingly of what they do. Souls in the heavens and on the Earth are working in unison energy, sometimes not knowing what they are working with, but dealing with it anyway.

You that work, either consciously or unconsciously with the field of Light, are Light bearers, working yourselves through that tunnel of darkness, none of you knowing where you are coming out, whether you will emerge on the Earth planes or in the heavens. It is individual and there is a time, a point, where the magnetic flow will take all souls where they need to be.

There will be many physical survivors, but there will be souls that are left for some reason or another to go on to different planes of existence. These planes of existence may be to other star systems. They may be melted into the higher consciousness Light. There is a myriad of different possibilities, but you can be confident that each soul will find its rightful place. You have always worked hand in hand with the fourth dimension. You have risen through your own experiences, through your own yearning to move on, through your persistence to gain Light. That persistence allowed you to climb the mountain, tired and weary though you have at times been. If some of you have not acquired this level of consciousness, then quite clearly there would be no survivors, because the energy to which your planet will be subject is so rarefied in terms of physical existence, it would not be possible if it were not for the work already done. We say to you, every consciousness that makes this journey, every soul that brings itself to the point where it melts into the heightened consciousness creates more strength, more vibrations to lift and to hold mankind in Light. There is no time to pontificate, to philosophise on what will be - the whys and wherefores. It is absolute reality. It will be a movement, a change beyond any change known on your planet.

There is no way of total understanding in words. The reality is beyond words. Like love, it is an energy frequency that just exists. It is at one with the universal Light, and it has no reality in a word and verbal fashion, but you have come to the point of understanding within the soul, within the heart, the greater heart, the greater love. That point of understanding is what you could perhaps call trust. Trust, for you, is a word that is emotive. It, in turn, depicts blind faith. It is neither of these things. Trust is the knowing, the absorption, the communion with the consciousness of Light.

We speak of communion of Light and we speak of communion of Light beings. All around your globe there is no individual that sees or even perceives them all. Even the lower levels of the deva chain are being called in to help with the movement. Every plane, every living force is working through this. The beings of Light are being joined by other beings of Light from other places. You do not stand alone. There are the beings of Light coming out of a kind of curiosity, a link with your planet from the past, or just because there is an open doorway through which they can walk, to link and communicate with your beings of Light. The doorway to other globes and other globes' consciousness is now well and truly open. It is no longer just a fantasy or a vague concept within the mind. It is a reality, and because of the need of humanity at this time, these Light beings and communion of White Brotherhood stand hand in hand to help. When we speak of help, it is not like a band, a group of beings whispering behind a veil. It is an automatic process. It is a fruitful acknowledgement of the advance that can come for them in their connection, in their communion with the energies around your globe.

The very heart of your globe is being cut into now. It is moving. It is in a way like an operation. It has reached its point of strength within the energy field that you have been working within over thousands and millions of years, and in one sense, the heart of your globe has come to its conclusion, its end. But it is not final, because through the many energies now working above and below your Earth, there is an operation to advance and to make the heart of your world live on, accommodating the different flow.

You find it so hard to realise that your small globe is in many ways a leading force in the Universe and where it is going other globes will follow. Think not of yourselves now. Do not think of your material planes. Take yourself into the cosmos. See the working as one, and whatever occurs on a physical plane, there is always a moving force, moving, clearing, expanding, accelerating. A force that cannot end.

Dear friend of Light, you already are working with the beings of Light, and they with you. Even before you had become conscious of that, there is a unison of work. Some of the energies are not aware what they are working with, who they are being helped by. Of course it does not matter, because the communion force is still there. All you need to do is open to the being of Light. Just make available that space for them to help you. They are already here, they are in this room, they are everywhere, helping wherever they can. There are no exercises for this. There are no notes of attunement other than the note of attunement that you have already worked towards, which is the heightened note of attunement of your soul. When that occurs there will be a conscious recognition that other beings are close to you. But they still are there, whether you recognise it or not.

The changes are now very much affecting the body. To alter the DNA, the cell structure within the body, there has already been some work done on a heightened vibrational level. To allow the human body to accommodate the new force - and again we speak in picture form - there have been surgeons experienced in the change of cell structure. This has worked through the etheric body as we spoke of today, raining down into and through the physical, and this is why you are seeing some disturbance. The etheric blueprint is making possible the actual physical change of cells within the body. This change of cells is essential for the body to go with you at this time of change.

There are a few cases of babies being born with the new DNA, but those cases are relatively few. Some babies are coming in with very low frequency now, and the reason behind this is because they need to have the very last remnant of karma experience. Some babies, few though there are, being born are accommodating the new energy. As you get more and more towards the total change, you will find there will be more and more babies born like this. The population now will soar. There will be a very great need within many women to conceive. This is to allow, finally, the last karmic experiences. There will be some abrupt ends to that physical existence, and this could be disturbing for you, if it were not for the reality that most of you have gone beyond the time when you feel that physical death is in some way evil. Seeing these deaths is not evil, and it is nothing to promote fear. Unfortunately some orthodox and fundamental beliefs will look upon this as the work of the devil. How wrong they will be, and how dangerous is their force to suggest that people hold on to the remnant of the astral darkness that is so close to the Earth plane. When that remnant of astral darkness encompasses and holds a being, it will implode, and it is for that reason that we have always spoken to you firmly about fear, not allowing it any space whatsoever.

You must remember that each individual, no matter how young they are, still possesses individual responsibility. Even if you are dealing with a baby a few days old, you are dealing with individual soul responsibility.

There is no set process for healing. It is a frequency rate. If you are at that frequency rate you help. You are there. You help and it works. With or without your conscious knowledge, with or without any elixirs, it just works. If you want to give substances to people as a psychological effect, then so be it, but know it does no more good than just being, and giving willingly.

Scientists will get their proof very soon, and it won't just be with one child, it will be with many, and it will not just be with children, it will be with adults. At first, when they get the information, they will think that they had less knowledge before. Therefore, they will not see immediately the relevance of their data. But as more and more cells move and change, they will have to face the reality that the cellular structure is moving. One case does not make the whole picture - there will be many.

Because the cells are, as it were, being moved, the cells within the body are being excited. It creates a frequency that hums. So therefore some of you are experiencing great pleasure in doing just that - humming and making a sound and tuning into a sound.

This is wonderful for you. It becomes joyous because you can touch outer and inner sounds and vibrational frequencies. But the sound itself, played to someone who is dead to this, would have no reaction at all. It can, in some cases, just lift the energy and help the individual by aligning the frequency. There is no one instrument, one sound, one medicine, one experience that you have that makes this whole picture complete. It is you. It is your energy. It is your compassion of energy between yourselves and the heightened consciousness that is creating the feeling, that is creating the leading force, that is creating, ultimately, the lantern that guides you.

We must end today by saying once again that this time is a time of joy. We have just spoken of sound, and we say to you that you can ring out your bells in jubilation of the occurrence, the reality now. In ringing out the bells you lift your hearts and your souls in expectation of the new world and, dear children of Light, you will not be disappointed.

IX. MAN THE MAGICIAN

Man is coming into his own power. He is becoming strong, resilient and free. He is assuming the position of a King or a Queen. He is taking his own sovereignty. This has enormous potential and enormous repercussions, even on its own level.

We watch you as you have grown. We watch your acceleration. We watch your spirit. It is a Lighting up process, because as man unfolds and dissolves the many layers, the many veils, the spirit within burns brighter and brighter, and it gets to a point where this brightness, this energy, is like a magic rod, a magic wand. You do not yet realise that this is just what it is, a magic wand, and it can make magic for you. We have touched upon a very important point; that man needs to watch his thoughts, because combined with the power that is now within, he is now invincible and can create his own reality.

We have hinted to you that it is now time that man brings true his dreams by looking at his dreams of the perfected life, the perfected Earth. Within the atmosphere prevailing at the moment he can achieve this. He can make this happen. There are a few higher consciousnesses in the body at the moment that are working towards accelerating mans' dreams, mans' possibilities.

Your globe is coming to a point where it is like a completely clean page. This means that you are like your fabled Adam and Eve, that man will be the first of the new race, and what you do will lay the foundation for the next thousand years. So man must dream well. He must create his own energy field through those dreams. By dreams, of course, we do not mean sleep dreams, we mean the inspiration, the desire for good. This energy, this desire to good is being accumulated by the heightened consciousness, and put into a basket, gathered up to create the energy force, to make the new world that you know is coming. There are times in the history of globes where this occurs. Man is given completely free rein, almost as though he is starting from scratch. Of course in starting from scratch he has to handle certain implements, certain possibilities, and they are not the same possibilities as when, thousands of years ago, you entered another of your new phases.

It is within the minds and hearts and the consciousness of man, the story of Noah. The story of the great evacuation of creatures is very close to your conscious mind at this time, because although we cannot give an absolute statement of what will occur, we can say that a few heightened consciousnesses will create the new world, will bring with them the best from the past, taking it through into the future. By this we mean the best ideas, the best thoughts, the best abilities, the best in man, the dreams of man to make good, to become perfect, to become powerful in Light - this is what will come over into the new world.

We speak of new world, and we have to be specific because some of you reading this pick up these sentences and words and imagine them to be what they are not. When we speak of the new world we are not speaking of a new physical globe. We are not speaking of a new land, We are speaking of the same globe, the same land, the Earth. What we are speaking of, is a new frequency, a new experience, a new place in as much as your Earth will be cleared, will be made good for you to start again. You do not start again. Because you have evolved, you are not starting from caveman, you are starting from man, man at the early stages of his spirituality and the union of spirituality with the physical. This is where you are. It is a very good starting point for the new world.

Again, we say to you it is much to your favour and to your benefit that you have chosen to walk round and to come again and again to experience to grow, to bring forth by every different incarnation one more bit of the puzzle, one more energy frequency. In some cases a long tedious life would have only given one little bit of glimmer, one little bit of hope. Yet that one little bit of hope took you on to a new life, and how much you have expanded and grown. You stand new and fresh on the threshold of your spiritual union with Light.

There has been some confusion over the reality that you can, by what you have done, change what is to be. On one level man has come many, many years and many, many experiences, through to where he is now. But there is a great cosmic wind blowing that is changing your planet. Your mother Earth would be transmuting whatever you have done. So it is your growth and the Earth's growth and the cosmic growth that are the considerations in this change and this moving point. One by one, as each man takes on his power, his dreams, his thoughts are becoming reality. At this point we say, once more, personality difficulties must be completely jettisoned. They are of no value.

When man stands at this point of growth catching sight of the other side, he thinks to himself "Well this is all very well, but where is the fun? Where is the joy in being in this oneness?". Part of the fun, man realises at this stage, has been in the struggle. What will there be if there is no struggle? How can he grow if there is no struggle? We can only assure you that you will grow, and that your growth will not stagnate when you move into Light. Indeed the rate of acceleration of growth will go beyond anything that we can project at this time, because the possibilities are immense.

You have within your music a scale of a certain amount of notes. It is as though you have developed and outworn the combination of those notes. We say to you, you have access to new notes. So the combination of the new notes with the old creates immense possibilities. Now can you see that man will not stagnate. He cannot. Nothing stagnates. If it does, it just goes back into the Earth. It is not possible for life to stagnate, because the very fact of life is growth. It moves always, ever forwards.

Man's consciousness now is expanding, and by the time these words go into print there will already be in your media and within the scientific fraternity a buzzing, under the surface, of knowledge that is permeating through very strongly into the consciousness.

Much work has been done on the higher planes to feed this new sphere of activity within the conscious mind of man on all levels. You will begin to see the permeation of this growth by the time this year is at an end. It is somewhat sad that at a time when your world is moving, and in many ways will alter completely, that man will come to fully understand what the Earth is and in what he has been living. At the end of this stage, this enormous stage that man has finally come to, he will be going through in full consciousness of what has been achieved and what has occurred.

There have been writings about judgement day, and clearly this has been cloaked and misinterpreted. But we say to you, you can talk of a time in the very near future where there will be an instant realisation of man, and of what he has been experiencing, not only man, living man, but man in spirit, the souls of man and the souls of men waiting for judgement day. They will be awakened and they will be enlightened. As they are enlightened, they will move over to the new framework, the new world existence and the new world order of which you dream.

There are many organisations, and connected to them are many guides and masters, impressing upon you, at this time, the possibilities of man. The words used are 'the new world order', 'the new world, the new growth'. All these groups are in position, not because they will have power in themselves, but because of the dreams and the inspiration that it gives people within the organisations to which they belong.

Even within the heavy world of matter and the largely corrupt atmosphere of the political worlds, there will be consciousness of the new world order and the dreams that these leaders will have to accommodate. It takes, now, a very special person to be a leader, and it is a hard way ahead for any such person in that position of authority. It is not an easy place to be, and for many it will be very uncomfortable indeed. They are there because it is right for them to be there and largely they recognise their own destiny.

We speak of the fruit of mankind, the fruitfulness, the joy and the love that has been given through man, through his conscious acceptance of love, through his conscious acceptance that the struggle is worthwhile. Very, very few souls gave up their struggle. In the whole of this huge period of man there have only been, in comparison, a handful of souls that gave up the struggle. When you consider what the struggle has meant, this indeed is an achievement, one in which we did not expect. So we stand, as it were, hand in hand with you now, appreciating your journey and appreciating with you that this particular struggle is coming to an end.

Now, fortunately man is truly beginning to understand that physical death has no real part to play. Physical death is only a point on the journey, a crossroads on the journey of the whole existence of spiritual matter. Also understanding his power in itself takes man forward. Understanding that his dreams are now a reality takes man forward. You are standing above the clouds looking down on a new reality, and the possibilities ahead of you are very, very vast. You are beginning to catch sight and know of possibilities beyond your wildest dreams.

Even as we speak to you now, man is dramatically losing his fear en masse. You will have to see a few more years of struggle, and that struggle, in some cases, may be immense as it dissolves the pockets of negativity around the world. Again we say to you, look at these pockets of difficulty within your globe and know with surety that this is the end of this kind of struggle, this is the end of this kind of evil, this is the end of this kind of negativity. This is the transmutation, finally, into Light.

It is essential that those that carry on into the new world have their dreams and their thoughts in good order. Work on your thoughts. You are linking now to the mind of God, and the mind of God has power. You instantaneously create your thoughts, how important to instantaneously create the best dreams in man.

There are so many little facets of information that we could impart, some of which would pull you away from the main purpose of these words. The main purpose being inspiration, the seeding of new thoughts, new growth.

With all the information we give you there will be just as many questions by just as many people. However the questions asked will all be answered. Reading this material can open up, within the mind, possibilities.

We speak to you now through a channeller who lives within the island off the main European land. This island has been a proud and positive land. It is, however, a land where the energies are imploding. It gives off an energy into the atmosphere that is beginning to act like a leading star. The energy of this land is going onward, creating in the atmosphere a positive leading energy, and as the energy moves onward, the physical state diminishes.

There will be changes of structures within your land mass within a very short period now. It would be a mistake for you to believe that those changes in specific places somehow mean that that particular place has been a place of darkness. In some cases, and this land we speak of now is one of them, it is quite the contrary. You must not mistake the change for darkness. Change is not darkness. Death is not darkness. Physical death is a transmutation into a greater state, a greater movement. When a baby is in the womb you do not say it must be in the womb forever. It has to change its state. It has to change from breathing through the mother to breathing on its own. You understand that change, that birth, that growth. So you must finally understand that physical death is such a thing. It is birth. It is growth. It is transmutation. It is necessary. It is beautiful and it is joyous. When man finally realises this, he has broken the pattern of his own evil and his own darkness and of course his own fear.

Man has a challenge ahead now. Although the fight between good and evil in your astral atmosphere has already been won, man has to live through the clearing-up process, and man must be strong, must catch hold of his power, must know his strengths. This is the challenge of years ahead.

There cannot be failure, that is not a possibility. But there is clearly a challenge for you, and if you do not meet this challenge your energy will survive. But is it not better for you to jump over the threshold rather than have to work your way round the karmic wheel once more? Each soul that manages to accept the power, the strength and the movement, helps all the others by the positivity of his or her own force.

We have spoken a little about what is occurring with the souls on the astral field at this time of change. We have imparted to you that the astral dimension will clear in your world. How does this leave your loved ones, your souls, many of which are still there within that astral place? We tell you that some will move on, or else incarnate in a new, young Earth that is, in terms of your distance, many Light years away. Energy is never wasted, nor can it be. Within the astral planes of existence the movement is just as positive, just as real and just as important as the movement of consciousness on the physical plane.

Work is being done to allow people to let go of their own fear, so that this information can come further into the consciousness. We can speak categorically of a movement, evacuation of souls at this time. For some this evacuation is physical also, but it is not so for every soul.

There is an enormous clearing occurring. This means that each soul will be drawn to the appropriate place. We have hinted and we will tell you now that there is a large evacuation of souls going to a new land that will not incarnate on the Earth again. They will go to a land that is not dissimilar, but is largely clearer than your Earth, that does have a similarity of karmic influences, but not the same. This evacuation of souls will give that particular land, that particular place opportunity for growth, just as millions of years ago the evacuation of souls came to your Earth to do just that.

Some of you know this to be so. Some of you even remember your first time here. The Earth will be, for a short period, unstable. It will be extremely difficult to live in a physical sense on your planet. There may be some that can transmute their energy enough to stay. Like Noah, evacuating some beings is going to be essential for the continuation of life on your planet.

Some of these souls and people who go physically will not even know they have gone. It will be to them as though they have blinked their eyes and come back. Some will go in full consciousness. You have to understand that each individual has their own capacity, and the soul works, in human terms, like a magnetism. It goes where it can go, it gets drawn where it can go. In some cases this will be without the body, and in some cases it will be with the body. No one must be alarmed at this information. It is a natural process. You must take it as a journey of growth.

You know, from the shores of your country, man journeyed to new lands. He went across to the Americas, down to your lands in the lower half of your hemisphere. He made new lives there for himself and his families. It is no more unusual than that. But man will take himself to a new place if that is what his soul desires. When we speak of soul desire we mean soul magnetism, and indeed the word 'desire' could and can be described as a magnetic force.

We could give you more information on this. Specific information is not required, nor is it necessary at this time, but rest assured, all of you, that man will find his place. He will be where he should be. He will grow and transmute into joyous acceptance of the new order of Light.

Meditation for some is essential. For others, who have found their strengths and are at one with themselves and with God, it is not essential at all. For those that still teeter on the brink it is one of the things that can help align to their true selves, their true Light. Some of you are in a permanent state of meditation and some of those don't even know it. Man must find his own intuition, and trust his own inner knowing, his judgement that comes from that deep soul level. Man must do this now, or else he will get left behind on this transmuting journey into Light.

We will leave you today with the thought that everything is perfectly as it should be. When you look at difficulties, starvation and wars, it is so easy for you to assume that all is wrong. Dear children of Light, rise above that fear, and know that all is well.

X. NO LIMITS

There is a gathering of souls, a stillness of spirit in your world. This stillness of spirit is acting as an anchor, a framework for others to be brought into, like a pool that is made ever bigger.

Outside this pool of stillness there is much clatter, movement, excitement of the worst kind. The two do not seem compatible in any way, and yet gradually, like the clouds drawing up the rain, the pool of stillness is allowing in all those it can reach it. It is like a giant magnetism of Light, and as each little droplet of energy, of soul-Light comes into the great pool of Light, it adds to and enhances the main. This pool of stillness is connected to what you call your Universal consciousness, your group consciousness, your world consciousness. So all those souls within that pool know what everyone else is doing and where they are. Even when the mind is unsteady and wants questions answered, it has the answers within the soul, within the combination of the Earth's consciousness. This consciousness is linking to the group energies, the Brotherhood, the planetary masters and guides.

We have touched upon this before, but we wish to talk a little bit more on this subject. These planetary guides, that have worked for so long within your sphere, are moving very close. In reality those souls within the pool are actually part of the energy now. They are not separate. There is still a sense of communicating and touching with them, but in reality they are one information, one song, the outer guides and the inner soul touching together in harmony. This is always the way forward, because when expansion comes it mingles with the heightened energy forces. It becomes one with your soul, is one with God, and will one day be absorbed into the whole - be part of everything. This will be a giant expansion of Light as the individuality of the soul comes back in union with the whole. It happens with a little working along the way to that completion. At this present time there is a mass union with the heightened guides, the master rays, and that is why you can call yourselves masters, you can call yourselves teachers, because that is what you are.

The work that many of you are doing now is the work that masters and teachers have done throughout your history, the difference being that in the past there were so few others that touched that ray, that touched that Light. There seemed to be a huge gap in consciousness between those that were the teachers and those that were the pupils. Now there is much less of a gap. This is nothing to do with saints or gurus. This is everything to do with you taking up your rightful sovereignty, your rightful power of union, union with the angelic forces, union with the master ray of Light.

Some of you are already there, and yet you still need to ask the questions, or at least you still feel that you must. Yet when you meet with the possibility of getting the questions answered, those questions pale into insignificance. There are indeed no longer questions to be asked because, in reality, you have the answers within your soul. This does not always mean that you have the answers in a clear mental state, although you can filter them through, if you so wish. The higher conceptual questions have the difficulty of formulation, as many of you know. Become one with the master ray. Become one with the planetary brotherhood of man. Take up your rightful sovereignty, your rightful position, your rightful leadership and your rightful power.

Of course your journey has no end, and from this position you can now begin to see and link to other energies of Light. As your energy expands, like a ripple in the giant pool, you connect with and catch sight of other energy frequencies, other planetary logos, other brotherhoods of which there are many.

There are relatively few that you can link with because of the difference in substance, the difference in the realities between the two. But as you move ever upwards there are ever more energies that you can touch and one day combine with, being in tune with their rays, being in tune with their brotherhood, being masters and guides on that particular energy force.

The despair sometimes in your doubt, your questioning, your lack of trust, the remnant of fear that somehow someone can harm you or affect you. When you are in this energy frequency there is no one that can harm you. There is no right or wrong. There is only truth with different perspectives.

There are so many of you working in Light. There are none of you that are wrong. There are those that are still embedded in their fear, and those ones are working through in spheres of what you call therapy. Leave them. Let them work that through. There is little you can do for those. Some of them will transcend and some of them will not. Caught within this spiral of fear, they need to work through and expand. They need to accelerate and move from it.

We speak particularly today of those of you that know that you can be part of the ever increasing pool of Light, of heightened consciousness, and we speak firmly to you. When you are in doubt there are questions that you feel that you have to ask. We cannot tell you not to ask questions, but what we can tell you is the answers are within you. You are free within this space. You are in a state of freedom, and yet some of you still believe you are trapped. You cannot be trapped when you have made that transition of Light.

Those of you who have worked as teachers with groups and in organisations will see a rapid movement within a very short space of time now. Groups are undergoing an enormous change. It will no longer be a state of leadership in terms of one leader and the rest following. We will explain. Within groups there will have to be someone who is the spokesperson. This spokesperson may not be of a higher frequency to the others, it is merely that that person does that particular job a little bit better than others. So when we speak in terms of leaders, it will not be so in terms of energy frequencies. However, for practical purposes there will always have to be a spokesperson.

You must work with your groups to get the cooperation of energy. You must work for this cooperation of Light. Let in the whole group acting as a leader and a pulling energy, a magnetic force such as we spoke of earlier in the session for others to join, to come in and to absorb the energy and to move where they have to be. It is very much a coming and going situation now. Souls come in refreshing, replenishing, being part of, getting their strengths, their rock within, moving on, going out, ever increasing the pool of energy. This is happening like wheels within wheels because in reality the consciousness is joined. When there is a need for it, these consciousnesses will have complete union of information, and there will be a time coming soon when those consciousnesses will instantaneously have the same message to move, to be where they should be, to do and to take up their work.

All of these consciousnesses, have a part to play, like a piece in a jigsaw, to help humanity move. The part they have to play could in reality take a few short minutes, an hour or two, one particular moment, one particular time that is needed, and this time is coming very soon. There will be a collection of souls, a collection of bodies that encase the soul to move on, to attract, to be one with an even greater consciousness of Light, expanding the whole planet in unison and in growth.

Watch your trees, watch them and listen to them. The older trees and even some of the younger ones have knowledge stored within their group consciousness. There is a consciousness, a high consciousness within trees. Look and link to the higher branches. From the high branches there is a frequency that draws down from the sky, like Lightning flashes of Light, and those trees are like old Father Time. They have the information that you will need about your planet. Many trees energies will implode back down into the Earth, to be born again in a different time in new Light. This is being done willingly by these trees, by this plant life, and the energies that have always worked with the plants and trees are becoming very conscious of the movement, and are helping the plants as their energy implode.

Your leaders energies are doing the same work to those that are still in fear, to those that energies are imploding and going back into the Earth, to be born again in different Light, in different mode, in different time. You must impress upon all souls that the soul itself is timeless. Therefore there can be no fear of physical change, because physical change is only a stepping stone on the evolutionary progress of the whole planetary system. This evolutionary progress moves on and on and on, ever moving. There must be no fear in the change of the structures and the change and the movement of the human being.

There is coming a time when the magnetism of each soul will find its place in this giant leap of humanity and the giant evolutionary leap of the cosmos.

Each little group of energies, each little gathering of souls will protect and will help and soften, as it were, the load for others. Even those that make a fuss now, even those who protest and appear alarmed at what is occurring, even those that may even scream in fear will have no fear. It will merely be a reaction from outer stimulus. The soul is being tuned, tuned and retuned into the acceptance of what is to be. That is how it will be and the remnant of fear that does exist will be lifted through the astral planes, will be cleared by astral beings above and below. Some of those that choose to go, to leave their physical body, will take with them Light, clearing the astral planes as they go, with a conscious unison with the astral beings of Light that are there to help in the total clearance of fear. Already this work is being done. It started a long time ago. It is accelerating now, and there are many of you who are conscious of this work. Again we say to you it is no coincidence that there are so many souls that have worked and are familiar with these energies of the astral and the golden ray beings that came to this planet many lifetimes ago. There is no coincidence in this union now. Even when the conscious mind is not completely clear, the souls mind is very clear and knows what it will do when the time comes.

Vibrational frequencies are changing. We spoke to you before of the dangers of what you call noise pollution. Your hearing, your perception of sound and the vibration of sound is altering. Sound is being used very valuably in a healing way. But it is not the only answer, because always and every time we come back to the real truth. It is in your being, in your energy, in your willingness to participate with this energy, to expand and to give out the Light that helps move others. These are things that you use to help and heal, and sound is one of them. It is but one instrument, one way, and certainly not the only way. The reason it works so effectively is because it gives the mind the opportunity to be still, and it is from this stillness of the mind that the soul can best do its work.

There is, if you can perceive it, a joyous union of Light bodies, and your soul is amongst that union. Your souls are connected with the Light beings, they are not separate.

Please discontinue the idea of separation. Your channeller is allowing these words to come through because of her conscious decision to unite with the heightened frequencies coming through her, and at this point it is her, it is her heightened connection. It is not correct to speak of these communions of information as being outside and separate. It is you. It is a part of you. It is a cooperation of Light, a union, a togetherness, a completeness of Light, and for all of you it is possible, all of you.

Please let go of the great misunderstanding that it is only somehow some perfected being to be united in Light. You are all perfected beings in your possibilities. You are all part of Light if you allow the veils of your fear and your doubt to be finally removed.

When you get information, do not talk about where it comes from, who it comes from, what it is. Know what it is. Know it is a union of Light force. Know it is a combination of you and God.

When we give out the information, we give it out like little drops of diamonds, little facets of different bits of information. These facets are polished by their communication. They are enhanced by the words, but the words are nothing unless you connect with your own inner union of Light. All of you who are Light workers do not need to worry about a particular date, a particular time, where you should be or what you will be doing. When the time is right you will instantaneously know. You will have giant communication. All you need to do is keep your Light shining bright without barriers, without fear.

The frequencies of sound are altering. You have your structure, your cords, your scales on your musical instruments. You speak of a note in connection with an energy, and it is so. You can unite the frequency through attuning yourself to that note. There are notes beyond your hearing, and the frequency of one note could create a frequency of another one. We try to make this simple for you because it is part of the harmonics of energy - each individual harmonising themselves through the sound and the note that they harmonise, to creates this harmonic sound, this harmonic frequency. When this occurs the energy within the body instantaneously aligns and connects with heightened energies outside the body. This, however, cannot be permanent and can only give that individual feeling for a short time. However, as the alignment and the knowledge and the feeling of the alignment is there, it can allow that individual to know, to feel the possibility of alignment. In feeling and seeing the glimmer of Light always, there is a real possibility that that individual can never hide from it.

Sound, as we have said, clears the mind, and works better for some than it does others. There is not so much need for different sounds and different notes for different parts of the body. It is more necessary to have the harmonic note of the individual as a whole. This harmonic note allows healing of a superior degree to take place. In some cases, if this is given, such is the difference in frequency, it creates immediate healing, immediate Light, that the individual's energy changes drastically. So the administrator of these frequencies must be very careful, because if in finding the individual's harmonics they are unbalanced, to put them in touch and attune them to their harmonic healing note could do the opposite of the healing, and could disturb the energy. It is pulling in so fast the unaligned energy that it jolts, and that very point of jolting can disturb.

So it is not for everyone to heal through sound, and it works best with those who are somewhat a little way along the path of their spiritual acknowledgement. Of course, when we speak of healing we are not being specific about the physical, the mental, the emotional. The sound helps the physical very well. It can also help the emotional. It can help the mental by allowing a break in the disturbance of the mind,but it can't very adequately completely heal the mental body. It can and does help heal the physical. This is because the sound of the frequencies align to the etheric body which is so connected to the physical. The waves of sound on your planet are not compatible always with the mental waves of sound. Of course all these things are individual, and it takes a very intuitive inspirational practitioner to do this. It is certainly not the work for everyone, and it is not work that people should readily dabble in.

Again we say these things are one way, one possibility you can use as a springboard. They are not the answer in themselves. The answer is always within the individual acknowledgement of Light, and the acceptance and the trust in the beingness of spirit.

Alcohol and those kind of substances work upon the body in different ways, depending on the body and the person, but it is not to do with the body that we really speak. There is a point of growth where the individual is so sensitised to energies that he cannot take these substances. Then the heightened consciousness takes place. They can come back to them if they want to. Of course, by that time it is rare that they do. It is not so much a harmful experience as an experience of no particular value.

We spoke earlier about alcohol and drugs when we were speaking about the connection with the astral forces. When an individual is working with those then he will, in some cases, find an alarming sensitivity that does create not a damaging affect, but an upsetting affect on the individual.

We have spoken a little about foods and the intakes of all substances. This is unique to every individual. There comes a point in the progression of each individual where they become, as it were, very sensitised to things. This is the point when souls decide they want to eat fresh food, good food. Very often they leave behind heavy foods such as meat. It is at that point that the sensitivity is such that they find alcohol and drugs a difficulty. Because of the major turning point in man, because so many souls are coming up and through the consciousness, you are seeing many souls in this sensitised state. But we have also said to you there will come a time when man will then be able to eat and drink anything including foods that are dirty and full of bacteria, because their body will be so strong and resilient. This, of course, is a heightened state, and most of you are not at this point. You are seeing so many souls looking to their diet, and it is right that they do at that time. They are sensitive to food and to drinks and to drugs. It will not always be so if they journey further into a consciousness state. Where the consciousness of the physical body aligns to the consciousness of the spiritual then that individual can and will be able to eat or drink anything.

When we spoke of the difficulty of drink and drugs we were aware that there are many souls in this sensitised state. In all cases it is better for the individual to know and judge for themselves what they need or want, and in the case of taking in any substance, the individual must be attuned to what they need. This they cannot learn from books, they cannot learn from anyone else, they must just be intuitive enough to realise it.

However, when an individual is at this sensitised state, to drink and to take drugs could waylay their development and growth, and could open them to a negative quality. But you will know for yourselves, you will see for yourselves. The kidney in man is altering now. What flows through the organs and how it flows is altering. The heart is altering. The liver is having to work overtime, and there will be some illnesses now connected through the liver. This is alright. This is the adjustment of man to his frequency. Again we say to you, realise that what is occurring and taking place is so rare and unusual, it is such a marvel, that spiritual energy is aligning with the physical.

There are two different energies. This is the polarity coming into one. These are frequencies and energies that, up to this time, have been rare in their union, extremely rare. Now man is going through this transition with both his spiritual and his physical - the alignment of the polarities. We cannot say what this means. We cannot say where this will take man. We only have a kind of projection. Man has his destiny in his hands and never more so than now as he comes into his own sovereign state of beingness, of oneness.

++++++++

The mental body is very disturbed at a stage of mental illness. The reasons for mental illness are long and innumerous. Basically, like attracts like, and the mind is very tied and connected with the soul. Sometimes you get a soul coming into incarnation that has had disturbance before, and he brings it back. This is different from emotional astral energy. We are speaking specifically on the mental and the mind level. Such is the magnetism here that it is often very difficult to get through to these people, and certainly if you try to get through to the mind or the logic you are wasting your breath. If you can bring down the heightened forces of clear mental forces through the mental body and work in that sphere through the guide energies of the mental planes, you can alleviate this disturbance around the mind and in the brain for a short period, allowing relaxation and allowing the magnetism to be released. Sound can help with this, but be careful if you use it, because you must choose your frequency carefully. You must accept always that the magnetism is with the permission of the individual. That individual has permitted, and in some cases wants that magnetism but is frightened to let it go. The mental body, the mental state has always been a difficulty for man. But now, as you move onwards through the lower states of physical emotional bodies, man is actually beginning to understand the frequencies and the possibilities of the mental planes.

The mental planes, like the astral planes, have energies working with them, and if you want to do this work you must establish your union with these beings. We speak of beings, but they are not beings in the same way as your spiritual astral guides are beings. They are more of what you would call a ray, an energy. Establish your connection with that ray, that energy, and those, as it were, salamander type energies will help you with your work. You must ask yourself however - and we speak with gentleness here - why these souls are so attached to their mental disturbance.

You must ask yourself, always, whether it is really possible to help some individuals. We have to say to you that sometimes it is not. However if you are intuitive enough to know when you really can do something for that soul, you must do it. If you call upon the mental rays you can stimulate, even in very large groups of people, heightened consciousnesses, and masters have done this for many a long year. You can send out that ray en masse, which will not just heal those who are mentally disturbed, but will align the frequency of the mental body to the other bodies within.

You must understand, dear souls, you have control. You can and you must trust your connection, but make your connection with the mental planes if you want to do this work. Create that stability. You are in control. The energy works through you, only through your desire, not through the desire of the rays, but through your desire, your will, your power. Understand that and know its value.

Think on the stars in your heavens. Think of the myriad patterns. Think of the enormity of your cosmos. Think of the many light years, the hugeness of the physical cosmos. Then realise the energies of the cosmos and the other possibilities of energies unseen which are every bit as much as those that are seen. Take your mind, take your inspiration, the enormity of the possibilities. The 'no limits' that you have because, dear children of Light, there never will be any limitations other than those that you create for yourself.

XI. SONG OF TRUTH

It is important for you all now to absorb the frequencies that are all around you, that are there for you. It is right that you do this. You need to adjust your frequency level by this absorption. All you need to do is to be conscious of this band of energy that is in position now, pulling up consciousness, raising levels of intuition and guidance. Indeed it is this very level of energy that is making possible the transmission that we give to you today. All you need to do is to be aware of it. Breath it in from the top of your head - the Golden Light Ray. The absorption of the energies of truth and heightened consciousness act as a rarefied oxygen stimulating your mind, your body and of course your soul.

The struggle is over, dear friends of Light. There cannot be and there must not be any more struggle. What will be will be. Go with it in the flow. This is not a state of giving up, it is not a state of despair. It is a state of completion, the end of a journey, the job well done. And the job has been well done on your planet. All you see, some of you, is disaster, is fear, is difficulty, within your own life, within the lives of others. This difficulty is to your benefit, is your power, is your growth. It has been successful in its enlightenment aspect to the world, to the cosmos and the Universe. In some ways now there is very little that we can give to you because you now come to this point of completion, and those that are not heightened in energy and who turn away from this information on some conscious level or another will go on to a different place, a different time, a different state than the ones that are moving on by the fluidity of the movement of Light.

We have spoken of a parting of the ways, and indeed the parting of the ways is starting now. There is a tiny amount of time when those that stand on the brink can jump, but it is very little now, and within this year it will be coming to an end. This is why the immediacy of this information is so important. It isn't of course the only information. This very information we give today has been permeating through the consciousness for many a long year. It has been given out in one format or another. It is not one voice, it is a universal soul's voice singing out the song of truth. If you do not jump now, the potential of the union of the heightened vibration of the fifth dimensional Light force will be over for you this time around. But the wheel always turns, always moves forwards. There can never be stagnation.

In the next few months you will hear of those within their rigid structure being frustrated and thwarted. These rigid structures particularly apply in political and scientific schools, and within science there will be a kind of war. There will be a parting of the ways of those that trust their intuition with nothing to back it up except theories and beliefs and knowledge within, and those that say this cannot be so, that you are wrong, you are delusioned, you are foolish. It is not so easy for those men and women in science and within the religious structures to go where they cannot see, to go only with their intuition, only with what they know within. It will be an enormous step of faith, but it will not be blind faith. Indeed those that do not take this step will be more blind, will pull up their cloak of darkness further up around their shoulders, and continue their life within their rigidity and their structure without any progress. It is true, within the religious schools and within the political schools, there is no structure, no hard and fast belief system or even scientific formula that will not be put into question by the faith and the truth, the intuitive truth within.

As the ways part, the heightened intuition will come into its rightful place, as it needs now to largely take over from the logical conscious mind. When we speak of taking over we do not mean that the conscious mind disappears, of course it cannot. The intuition is the heightened energy that needs to, if you like, amalgamate with the conscious mind, and as it blends with it, without fear, that intuitive being at last becomes at one with the reality of the connection with Light.

Much of your struggle has taken place because of the separation between the conscious mind, the so called reality, and the heightened, what you call intuitive and unseen forces. Those now mingle in unison, and those that have already accommodated this will wonder why their conscious mind does not work in the way it once did. They will wonder why they may be forgetful. They may react differently, or not react at all. Their conscious perception has altered tremendously in this change. It feels, initially, like giving up, like you are out of synchronization, you are out of true reality. But, dear children of Light, trust your intuitive knowledge, the heightened intuitive knowledge that is part of your soul conceptual instincts. Trust that, and you will see, you will know that your reality has altered as your conception completes itself. With this change of perception the world will change automatically.

We wish to speak on desire now. Those that make this leap into consciousness now must finally recognise that the desire is part of this reality. So many of you have been wrongly taught that desire is wrong, and if you desired something you must push it away. You were told it is wrong to desire, it is wrong to want. If you have felt desire in the past, you have sacrificed what you really wanted because of these attitudes. Desire has always been part of the journey. Now desire becomes complete with your unison Light, and you must trust your desires are one and attuned to exactly what is right. Because as we have spoken to you before, you cannot go wrong when you are aligned to Light.

So trust your desire. Trust it and know that it takes you along the path of absolute and ultimate truth and ultimate reality. Of course when we speak of ultimate, we do not mean a finishing point. Again we say to you there is no end. There never was a beginning. There was a beginning of the physical worlds, but there never was a beginning of the soul Light force. It is hard for you to comprehend that the Light force that exists within you, that which you call your soul, has existed for time immortal, and will exist forever. It can never be destroyed. Time itself was created out of the necessity of the physical experience. Time does not exist in the reality of Light.

Many of you are becoming aware of that knowledge, that consciousness, truth and the Light force is within. And so, as you take your stand one way or another, you hop on, as it were, another wheel, another state, another perception. You join with what we describe as the dimension of Light. That wheel will take you on its own journey, on its own experience. It is large, it will turn around and around as your karmic wheel has turned, until it comes to its point of completion, and then you will transmute again, and again, still keeping the physical body, until that point of total completion of the physical worlds when you will then melt into a completely different reality, which we cannot even speak, because it is beyond your comprehension and your knowledge. As you make this momentous connection and journey into Light, you melt away from the struggle and the karmic difficulties that have been so rich in experience for you. This is primarily why it is such a major step at this time.

We always say to you listen. Of all the things we have ever said, this is the most important and urgent for you to understand and acknowledge. To listen to that soul sound, to listen, to hear, to trust that heightened intuitive quality, to trust your desires as your desires complete with that energy. If you deny what you hear within, you are denying the very journey into Light that you all ultimately desire. That is the truest desire, the highest force, the highest want and connection of the human being. The desire to do good has always been very strong within you, and that desire, that attunement to good has led you many places with a richness of experience. How foolish it was of those that told you in the past that your desires were evil, because the richness of that desire to good rings so strong, it would always take you automatically where you should be.

That desire to good is still within you, all of you, within those that jump into the dimension of Light and those that don't. Desire to good is the predominant force within all of you, and because of that it was never really feasible for evil to win. There have been times when the evil permeated and the dark forces were very close to your planet. There were times of great darkness, great despair, great struggle, and perhaps none more so than what you have experienced, some of you in your lifetime, with the oppression of the racial disharmony in the Second World War. But even within that dark period of history and all dark periods of time the will to good, the desire to good, even in the dark holes of the prisons and the oppressive states, was always present, always strong. Within the very torturer there is good.

When there is recognition of the will to good as part of your strength and hope, it expands and evil diminishes. Finally desire and the will to good combine. It is the resurrection into Light. The resurrection from evil to good.

Now, after this year, you will truly see this parting of the ways. You will see the unfolding of energy within a mass that do not take the step. You will see the energy within this like a leaf that has been plucked from a tree. But do not fear for these souls, they have their protection from the heightened force, and there are those beings of Light within the higher astral planes, the angelic beings, that have elected to help those beings into their journey and their furtherment of their soul. So do not fear for them because they will be safe. They will continue and they will grow. It will be as though one half of the coin is turning, and part of the coin that is turned upwards is within those of you that hold the torch of Light. You will be the torch bearers. You will be the leaders. You will be the strength for the planet itself. Because when this occurs, before the middle of next year, there will be, for those beings of Light elect, a tremendous connection to the Earth's consciousness. You will automatically help what you call your Mother Earth in her transition, in her consciousness change, and indeed this will be the work for the future, or at least a part of it, to make this conscious union, to be at last in true sympathy with the Earth's quality.

By your leading torch energy you will help the Earth change its polarity without fear. The Earth has its own awareness of this, and it is concerned in its own way, as its magnetic pull goes, pulls, goes back. The needle of its magnetic pull is fluctuating. You, dear children of Light, will have that marvellous union at last with the Earth. There have been points when you have been connected to the Earth and some cultures, some qualities have been more aware of this than others, but you have never been part of its consciousness. Very soon you will be able to do this, and those of you that absorb this consciousness fully will in the future be able to be a planet, be able to be a new globe of the future. Think on that! There will be those of you that by your Light force will, by its unison consciousness, create a new world for new beings in a new time.

There is an opening in the cosmos for this new growth. There will be new planets born now, and the consciousness that you make will help. Those that do will be like Old Father Time. They will be there for the eternity of that particular planet, which could be many billions of years. So when you think of living forever, think on the Light force of the planet that lives not just a handful of years, not for just a decade, but for billions of years. Think on that!

Your thoughts, your desires and your desire to good is so very important now, because through your desires, through your dreams, new worlds will be created, and the planet on which you exist will be able to turn into its true destiny - the planet of Light.

We try always to paint you pictures, to give you analogies. But we can only say to you now that if individuals recognise the change and do want the change, and have desire to good, but are still trapped within their personality fears, so be it. The opportunity is there. There is an opportunity in life to experience one thing or another. In the higher scheme of things everything works out its own path. Every energy gets pulled to its own place. Fear will be dissolved completely to all beings on your planet now. Those that cannot make this conscious union at this point will have their fear dissolved, but will have to experience some more years in what is similar to a karmic experience. So they will live out their lives such as it is at this time, and when they die, which will be in a multitude of ways, they will not come back to planet Earth, they will go to one of a number of suitable planets. There are many opportunities for those that do not take this leap. Their opportunities will be in a kind of spiritual plane. It is not like your astral

114

plane, but it is similar. There are opportunities to help other planets that are coming up in the evolutionary pattern towards yours. Indeed those that see the Light and are aware of it, but do not make this leap, will be very useful in other spheres, because by making this connection now they can help those in another time, another place, and lead them into their completion. Within the scales of time it is huge, and in truth they may have to wait for thousands of years for a further task and a further reincarnation. However time is nothing, and they will feel nothing in terms of soul existence.

We speak to you very kindly, very gently, when we say there are some that are so low in their energy force that their energy will go into your Earth and they will not reincarnate as individual souls. They will go into a mass that will help the Earth with its energy and its alignment. There will be those now that cannot reincarnate in individual consciousness. Those, anyway, have no conscious awareness of what we speak.

Fear has always been your greatest enemy, and yet fear has helped you, because through your fears you have journeyed far. The very fact that you have pushed through your fear, even when it has been intense, has excelled your populace beyond your comprehension. So do not see fear as the great devil. Indeed we ask you to see fear as the grand awakener, and in that knowledge you can throw off, finally, the cloak of its hold over you.

We give the star image of Light. Take it in. And again we say to you all, absorb the Light and absorb the energy, it is there for you now.

XII. THE COSMIC ADVENTURE

The unison force of Light beings surrounds you to unify and cement all that we have given you. This work is not ending. We continue and expand, but in this guise, in this format, for the time being anyway, it will pause.

We have tried to impress upon all of you that the changes occurring are not evil, they are not destructive. They are enlightenment, to do with strengths, to do with enjoyment and recognition of the spiritual forces and powers. This has been our major work with this channeller, to break down finally and irrevocably the fear, the tension and the desire to hold back when all of you must go forward now. This is what we have endeavoured to do, and already there has been some little success.

When we speak of death, immediately some of you imagine this to be a terrible occurrence. But we must speak to you now of death, but we say beforehand that we speak about death as it is, a joyous rebirth of spirit.

There will be many sudden deaths, individually and en masse. Do not fear, this is nothing to be alarmed at. This is part of the process. Hold fast to what you know. There are ups and downs to come. Your strength is needed and it is also assured, because those of you that have made this commitment to Light will be made like rock. This does not mean that you won't get disturbances of the physical body from time to time, but these will wash over as normal and they won't affect the rod-like force within.

Again we try to impress upon you - and we think carefully of the words that are appropriate in allowing you to realise - this is an amazing time. It is the end of the season. When the leaves fall off the trees they are dying, but you know they will come again in Spring. The Autumn is very beautiful. It is full of rich colours. It is the harvest time, and so it is with the old energies. The old energies are harvested. Their nourishment has been given, and the trees lay bare for a season, to be born again. It is in this time when the trees are bare that you would be forgiven for thinking all is wrong. All is right, dear children of Light. There needs to be Winter for Spring to come again. In reality, even when you do not see the leaves on the trees, the sap is beginning to rise. There is very, very little time when there is no action. Nature dies down for a very short period, stops, and then immediately comes full circle. This is what is occurring. You have come full circle. You will move up the spiral into a place where you have never been.

We started by explaining to you that the change that is occurring is in three separate ways, a change of season, if you like, in three places; on your globe, in your system and in the cosmos. We speak again of wheels within wheels. Realise your strengths and your communion with Light. Understand now that realisation must take place. If it does not your energy will implode.

Take up the banner of Light. Know completely the heightened consciousness that you feel is joined to the heightened consciousness of your planet, the solar system and the cosmos. It is for this reason that so many of you now are experiencing the different layers and levels of Light beings. Some of you are communicating with the Angels, with the spirits, Earth spirits, with the beings of Light from other globes from all over the place. Indeed the veil between the layers is very thin.

As this energy of heightened consciousness permeates through the atmosphere like heavy golden rains, it has enormous affect in ways and places you will not see.

Those that waste time on good causes are fooling themselves. The good cause is within. The right cause is the 'God force' within you. You must realise this and you must accept it.

Finally we ask you not to waste your time following paths that in reality will lead you nowhere. Of course everything done, with a note of purity, a resonance of Light is always powerful, always productive. We ask you all to join together in your mind, in your thoughts and in your being with the Light above your Earth, the heightened consciousness force. Because even those that in reality are not at that frequency can, with their feelings or their mental state, be joined. So it has the effect of pulling them and others with them. It is again like a magnetic force. The magnetism of this energy is getting stronger by the minute. Its pull and its force is becoming immense. We speak of it in terms of heightened consciousness, because this is a word you will understand. But it is not absolutely as it sounds. It is a combination of forces. It is most attuned to the mental force. It is a combined force of pure consciousness, pure will, and on the human level there is nothing as powerful as pure will, because pure will joins with the God force. This is what we tried to explain to you before in terms of desire becoming joined to the will of God. When your pure will is adjoined and adjusted with your desire, your desire is always in tune with the right and heightened forces. Do not waste time now. Join your will to that will of God. It is more possible than it has ever been. What was only open in the past for a very few is open now to many, many, many more.

The wave of this energy is pulling you in. Think of the wave of your sea, ebbing and flowing, pulling, sucking you into the ocean of forces of Light, because this consciousness is like an ocean. It is an enormous pool of energy. This became intact completely last year and it is growing stronger. In some ways the process has accelerated more than we could know, but the difficulty with one aspect of acceleration is it always needs the counter-reaction in the lower energy forces. This is why the lower energies of the physical body have so much been shaken. This heightened Light resonates, vibrates so strongly now that it has had to accommodate the weight of the physical body and the Earth itself.

Even now you have not realised what connection you have always had to your planet. Your bodies are connected to the force beneath your feet. So the heightened consciousness needs to enter into the very heart of the Earth, because it is struggling in its process of transformation. Some of you have sensed this and all to the good.

As much as you are giving healing to the Earth, you must give healing to your bodies. It is not right that you discontinue to look at your physical body. Indeed, we ask you to look at your bodies even more, because your bodies need the calming and the peaceful rays of comfort, because they do not have the benefit of the consciousness awareness of the radiance of Light. But they will shine out, and when they do the skin will become translucent. The cells will alter and it will be noted physically. Science will think it has discovered something new within the body. It has not. It is finding something that has always been there, but it is transmuting. So, look after your bodies now. Do not abandon them in their time of need.

The mental energy forces are adjusting beautifully. The emotional forces of darkness are leaving rapidly. The physical body must transmute. We cannot wait for further generations, because the mothers of the children to be need to have the new frequencies, otherwise the children will not survive.

Everything must come down, through the layers, into the dense physical. But this does not mean the physical is somehow pushed aside or is less important. The physical body, the physical Earth is very important. Do not ignore that fact. Enjoy your bodies as they bloom into new life.

We have spoken of the spiritual coming of age of all, the many layers and levels that have to run in unison, to be adjusted, to be aligned. So for some of you who are mentally strong but are physically weak, if you cannot adjust to that frequency, unfortunately this time round, this turn of the wheel, you will not transmute into a fifth dimensional being. Do not concern yourself. Energy, life force goes on. At different times it has needed different energies. We would not ask for all the physical beings and planets and stars to be of the same energy. It is right that there is a different energy process. It is right that there are different stages of evolution. It adds to the pattern and the wave of the cosmos. Accept your being of Light, and if you do not make this transition you will still continue to move and to teach and to help others if you can.

We spoke once of the reality that each age needs a teaching energy for that age. Each energy needs a teaching force for its own particular master ray. All the way along the evolutionary process there are teachers. There are students, and the students are the teachers, and the teachers are the students, and so it goes on. There is no start or finish. There is no failure or success. There is opportunity. There is enlightenment. There is enrichment on all the many rungs of the ladder. All those things are possible wherever you stand, whatever you are there cannot be failure. Remember that there is, however, opportunity and growth, new growth. Spring coming from Winter.

Ironically, some of you already have gone beyond need for this information. You do not seek it, you do not want it. Accept your alignment. Accept where you are and where you will be going. It is right.

You have thought of your mental energy as part of your thinking process. It is not. The mental forces of Light we speak of do indeed house thoughts, but it is a frequency of its own. It is not the thoughts that create the frequency, it is rather that the thoughts can be housed by the frequency. Mental energy is a refined energy and the highest level that the physical body can manifest. This is why, when we speak of the connection with the mind of God, it is referring to the higher state of man. Of course when you cease to be man you go beyond that, and beyond that and beyond that. There is no beginning and no end.

And so we bring you back to the cosmic wheel, the forces revolving in the whole Universe, the ebbing, the flowing, the cosmic breath and the acknowledgement that from the tiniest seed, the tiniest insect in the ground and the biggest star there is a connection, and that you all journey together. Indeed, we will say all of us journey together in this giant adventure that you call life

Free will is constantly in operation, but by the magnetic forces of the revolving wheel of energy you can, by your will, be part of one wheel or another. You can grow within one, move on to another, and this is what the dimensions are, wheels within wheels.

There is an enormous harmony in the cosmos that you cannot experience, and yet some of you are catching a reflective sight of the harmony that exists with every movement. There is even harmony within the growth and the dying of stars. We speak again of the seasons, the coming and goings, the movement from Winter to Spring. These are conceptual thoughts for you, and yet these thoughts have a momentum.

The sound of the Universe rings out with its vibration keeping in balance, keeping in harmony, all those spheres and globes and beings within those spheres and globes. Even now none of you realise how many living beings there are within your globe, or how many exist within the whole Universe.

Now the cosmos itself is moving through its own veil. It is coming completely into a different phase. This is too big for you to comprehend, but it is as though an energy has seeped in at the edge of the cosmos, moving and being absorbed like a sponge, creating a new colour, a new frequency, a new sound through the whole of the cosmos. When we speak of whole, we of course mean the expanding Universe. It is expanding and contracting. But even when it is contracting it is expanding. It never stops moving. It never stops its expansion. Now, almost inadvertently, it has caught hold of a frequency force which is being drawn in. This is altering many patterns, many forces, breaking down barriers, breaking down dimensions. This means that every living soul that is in tune with its consciousness force, that we have spoken of earlier, can create something new, something bold within that cosmos, a new star system, a new planet, a new moon, a new energy. Think carefully. Be wise. There will be no turning back, because the cement, as it were, will dry, and you will be where you project. This is coming very soon for those with heightened consciousness.

Never before has this occurred in such an enormous fashion. There have been odd shifts, occurrences, but it has happened in minor ways. Indeed your mythology is full of it. This is why the mental force energy is so important and that the thought patterns must be cleared, not just for those of you with heightened consciousness, but for all of you, so that there is a clear way for the projection, for the true desires of growth and creation.

These are enormous concepts and thoughts, but they are a reality. In some ways we could go on channelling for many more hours. There is always more to say in terms of what is occurring, but we have given projections, we have given you conceptual thoughts, we have sown the seeds. It is up to you to nurture these seeds of goodness, the seeds of energy within these words, to make your own pattern, to make your own flower to bloom if you want to, to fade if you do not care.

All of you, every living being, the tiniest insect, the huge planet, the biggest star, everything is cherished, everything is loved, everything is part of the whole, everything has something to say, something to teach and something to learn. There is not one thing in the Universe that does not have something to do or say.

Many of the saints and masters of your world are finishing their journey with you now. They take this opportunity to bid you a fond farewell. They have their own journey to take now, their journey of love and Light. They have served you extremely well. They have given totally, unconditionally, by the power of the love frequency of the master rays. Now, the younger forces, but the more potent life forces, Light forces, join to the love that they have learned from their masters, to enhance your journey, to guide your journey, your own transition, your own transmutation into your own Light. The Masters of Light take this one last opportunity to fill you full of their force and of their strengths that they have willingly given you.

We too now come to the end of our transmission. There will be questions and they will be answered, but they will be answered within your own hearts. Listen for the answer. It will always be there. You are never alone. With that acknowledgement, the unison of Light joins with you all in embracing you in joy and in love beyond measure.

Walk on your journey now, dear children of Light, in peace. You have much to be proud of, but there is still more to come. Never forget that you are one with everything. Never forget that we are all together in this adventure - the cosmic adventure of wisdom and truth.